Solitons: an Introduction

P. G. DRAZIN

Professor of Applied Mathematics, University of Bristol

R. S. JOHNSON

Senior Lecturer in Applied Mathematics
University of Newcastle upon Tyne

CAMBRIDGE UNIVERSITY PRESS
Cambridge
New York   New Rochelle
Melbourne   Sydney

Published by the Press Syndicate of the University of Cambridge
The Pitt Building, Trumpington Street, Cambridge CB2 1RP
32 East 57th Street, New York, NY 10022, USA
10 Stamford Road, Oakleigh, Melbourne 3166, Australia

© Cambridge University Press 1989

First published 1989

Printed in Great Britain at the University Press, Cambridge

British Library cataloguing in publication data

Drazin, P.G.
Solitons.
1. Solitons
I. Title   II. Johnson, R.S.
515.3′53

Library of Congress cataloguing in publication data
Drazin, P.G.
Solitons: an introduction/P.G. Drazin and R.S. Johnson.
p.   cm. – (Cambridge texts in applied mathematics)
Includes bibliographies and indexes.
ISBN 0 521 33389 X.   ISBN 0 521 33655 4 (paperback)
1. Solitons.   I. Johnson, R.S.   II. Title.   III. Series.
QA927.D72 1988
515.3′55—dc 19   88-1083 CIP

ISBN 0521 33389 X hard covers
ISBN 0521 33655 4 paperback

TM

To Judith and Rosalind

Solitons: an Introduction

Contents

Preface

The theory of solitons is attractive and exciting; it brings together many branches of mathematics, some of which touch on deep ideas. Several of its aspects are amazing and beautiful; we shall present some of them in this book. The theory is, nevertheless, related to even more areas of mathematics, and has even more applications to the physical sciences, than the number which are included here. It has an interesting history and a promising future. Indeed, the work of Kruskal and his associates which gave us the 'inverse scattering transform' – a grand title for soliton theory – is a major achievement of twentieth-century mathematics. Their work was stimulated by a physical problem together with some surprising computational results. This is a classic example of how numerical results lead to the development of new mathematics, just as observational and experimental results have done since the time of Archimedes.

This book has grown out of *Solitons* written by one of us (PGD). That book originated from lectures given to final-year mathematics honours students at the University of Bristol. Much of the material in this version has also been used as the basis for an introductory course on inverse scattering theory given to MSc students at the University of Newcastle upon Tyne. In both courses the aim was to present the essence of inverse scattering clearly, rather than to develop the theory rigorously and completely. That is also the overall aim of this book. It is intended for senior undergraduate students, and postgraduate students, in physics, chemistry, and engineering, as well as mathematics. The book will also help specialists in these and other fields to learn the theory of solitons. However, the theory is not taken as far as the rapidly advancing frontiers of research.

This book introduces the fundamental ideas underlying the inverse scattering transform from the point of view of a course of advanced calculus or the methods of mathematical physics. Some knowledge of the elements of the theories of linear waves, partial differential equations, Fourier integrals, the calculus of variations, Sturm–Liouville theory and the hypergeometric function, but little more, is assumed. Also, some familiarity with the main ingredients of the theories of water waves, continuous groups, elliptic functions and Hilbert spaces will be useful, but is not essential. The relevant ideas from one-dimensional wave mechanics (both scattering and inverse scattering), necessary for the presentation of the inverse scattering transform, are described. References are given in the text (or at the end of each chapter) to help readers to learn more of the foregoing topics. Some of the diverse applications of the theory of solitons are mentioned only briefly, either in the main text, or in the exercises at the end of each chapter. However, the Korteweg–de Vries equation is derived for a water-wave problem.

The material is presented as simply as we can, and a number of worked examples are also used to help the reader follow the various ideas. Of course, some parts of the theory are more exacting than others, and some problems are more difficult than others. The more difficult sections, paragraphs and set problems are indicated by asterisks; these passages may be omitted on a first reading of the book. Further reading is offered at the end of each chapter to direct the reader to more detailed treatments of some of the topics. The sections are numbered according to the decimal system, and the equations are numbered according to the chapter in which they appear, e.g. equation (1.2) is equation 2 of Chap. 1. The problems are similarly numbered (e.g. Q1.2), as are the answers (e.g. A1.2) at the end of the book.

We are grateful to Miss Sarah Trickett (Figs. 4.5, 4.7, 4.8), Mr Mark Lewy (Fig. 8.1), Dr Adam Wheeler (Fig. 8.2), Mr Gregory Jones (Figs. 8.3, 8.4, 8.6, 8.7) and Dr Stephen Thompson (Figs. 8.8, 8.9) for their computations and plots of solutions on which our figures have been based; to Miss Carolyn Pharoah and Miss Alison Davies for their clear draughtsmanship of the figures; to Professor Neil Freeman (various points) and Dr Andrew Wathen (§7.3) for technical advice; to Academic Press (copyright of Figs. 8.8, 8.9); and to Mrs Heather Bliss, Mrs Hilary Cartwright and Mrs Nancy Thorp for their careful and cheerful typing of the text.

Bristol   PGD
Newcastle upon Tyne   RSJ

The Korteweg–de Vries equation

1.1 Preliminaries

Wave phenomena abound in mathematical physics, and are met early in undergraduate courses. They may be first introduced as waves on a string, or perhaps on the surface of water, or in a stretched membrane. With a little more background they may be discussed in connection with sound – and then shock waves may be mentioned, or (for the physics student) the first meeting might be via electromagnetic waves. In all these areas it is common practice to develop the concepts of wave propagation from the simplest – albeit idealised – model for one-dimensional motion,

$$\frac{\partial^2 u}{\partial t^2} - c^2 \frac{\partial^2 u}{\partial x^2} = 0, \tag{1.1}$$

where $u(x, t)$ is the amplitude of the wave and c is a positive constant. This equation has a simple and well-known general solution, expressed in terms of *characteristic variables* $(x \pm ct)$ as

$$u(x, t) = f(x - ct) + g(x + ct), \tag{1.2}$$

where f and g are arbitrary functions. (In keeping with the usual convention, we shall regard t as a time coordinate and x as a spatial coordinate, although in equation (1.1) these are readily interchangeable since they differ only by the 'scaling' factor, c.)

The functions f and g (not necessarily differentiable) can be determined from, for example, initial data $u(x, 0)$, $(\partial u/\partial t)(x, 0)$. The solution (1.2), usually referred to as *d'Alembert's solution*, then describes two distinct waves: one of which moves to the left and one to the right, both at the speed c. These waves do not interact with themselves nor with each other; this is a consequence of equation's (1.1) being *linear*, and hence solutions of the equation may be added (or *superposed*). Furthermore, the waves described by (1.2) do not change their shape as they propagate. This is easily verified if we consider one of the wave components – f say – and choose a new coordinate which is moving with this wave, $\xi = x - ct$. Then $f = f(\xi)$ and this does not change as x and t change, at fixed ξ. In other words, the shape given by $f(x)$ at $t = 0$ is exactly the same at later times but shifted to the right by an amount ct.

Before we develop some further elementary properties of waves, it is convenient first to restrict ourselves to waves which propagate in only one direction. It is clear that this is an allowable choice in solution (1.2): merely set $g \equiv 0$, for example. A more practical approach is to set-up initial data on bounded (*compact*) support, and then after a finite time the two wave components f and g will move apart and no longer overlap. Since they never interact, we can now follow one of them and ignore the other. To be more specific, we may restrict the discussion to the solution of

$$u_t + c u_x = 0, \qquad (1.3)$$

where we have introduced the short-hand notation for partial derivatives. The general solution of equation (1.3) is

$$u(x, t) = f(x - ct),$$

where f is an arbitrary function and, since we could redefine t as t/c, we may just as well set $c = 1$:

$$\text{if} \qquad u_t + u_x = 0 \qquad \text{then} \qquad u(x, t) = f(x - t).$$

(We may also note the connection with equation (1.1): the operator can be factorised, and either factor may be zero,

$$\left(\frac{\partial}{\partial t} \mp c \frac{\partial}{\partial x} \right) \left(\frac{\partial}{\partial t} \pm c \frac{\partial}{\partial x} \right) u \equiv \left(\frac{\partial^2}{\partial t^2} - c^2 \frac{\partial^2}{\partial x^2} \right) u = 0,$$

where the signs are ordered vertically.)

When wave equations are derived from some underlying physical principles (or from more general governing equations), certain simplifying assumptions are made: in extreme cases we might derive equation (1.1) or (1.3). However, if the assumptions are less extreme, we might obtain equations which retain more of the physical detail: for example, wave dispersion or dissipation, or nonlinearity. Consider, first, the equation

$$u_t + u_x + u_{xxx} = 0, \qquad (1.4)$$

which is the simplest *dispersive* wave equation. To see this let us examine the form of the *harmonic* wave solution

$$u(x, t) = e^{i(kx - \omega t)}. \qquad (1.5)$$

(We can always choose to take the real or imaginary part, or form $A e^{i(kx - \omega t)} + \text{complex conjugate}$, where A is a complex constant.) Now (1.5) is a solution of equation (1.4) if

$$\omega = k - k^3; \tag{1.6}$$

this is the *dispersion relation* which determines $\omega(k)$ for given k. Here, k is the *wave number* (taken to be real so that solution (1.5) is certainly oscillatory at $t = 0$) and ω is the *frequency*. From (1.5) we see that

$$kx - \omega t = k\{x - (1 - k^2)t\},$$

and so solution (1.5), with condition (1.6), describes a wave which propagates at the velocity

$$c = \frac{\omega}{k} = 1 - k^2, \tag{1.7}$$

which is a function of k. (Note that c changes sign across $k = \pm 1$.) In other words, waves of different wave number propagate at different velocities: this is the characteristic of a *dispersive wave*. Thus a single wave profile which can be represented, let us suppose, by the sum of just two components each like solution (1.5) will change its shape as time evolves by virtue of the different velocities of the two components. However, this interpretation is virtually a repetition of the solution (1.2) of the classical wave equation. To extend the idea we need only add as many components as we desire or, for greater generality, integrate over all k to yield

$$u(x, t) = \int_{-\infty}^{\infty} A(k)e^{i\{kx - \omega(k)t\}} \, dk, \tag{1.8}$$

for some given $A(k)$. (Note that $A(k)$ is essentially the Fourier transform of $u(x, 0)$.) The overall effect is to produce a wave profile which changes its shape as it moves; in fact, since different components travel at different velocities the profile will necessarily spread out or *disperse*.

The velocity of an individual wave component is given by equation (1.7), and is usually termed the *phase velocity*. It is clear that equation (1.6) will admit another velocity defined by

$$c_g = \frac{d\omega}{dk} = 1 - 3k^2;$$

this is the *group velocity*, which determines the velocity of a *wave packet* (see Fig. 1.1). For many (but *not* all) realistic wave motions it turns out that

$$c_g \leqslant c$$

and, furthermore, the group velocity is the velocity of propagation of energy.

Thus far we have tacitly assumed that $\omega(k)$, the dispersion function, is real for real k. However this remains true only if we add to equation (1.4) *odd* derivatives of u with respect to x. If we choose to use *even* derivatives, taking for example

$$u_t + u_x - u_{xx} = 0, \tag{1.9}$$

then the picture is quite different. From equations (1.5) and (1.9) we obtain

$$\omega = k - ik^2,$$

and so

$$u(x, t) = \exp\{-k^2 t + ik(x - t)\} \tag{1.10}$$

is a solution of equation (1.9). This describes a wave which propagates at a speed of unity for all k but which also decays exponentially for any real $k\,(\neq 0)$ as $t \to +\infty$. (Note that the sign of the term u_{xx} is important.) The decay exhibited in solution (1.10) is usually called *dissipation*. Clearly we could have equations, like (1.4) or (1.9), which incorporate linear combinations of even and odd derivatives. In this case the harmonic wave solution may be both dispersive and dissipative (at least for suitable signs of the even terms).

Finally, let us briefly look at one rather more involved aspect of wave motion, namely that of *nonlinearity*. Most wave equations (like (1.1) and (1.3)) are valid only for sufficiently small amplitudes. If some account is taken of the amplitude (in a 'better approximation') we might obtain the nonlinear partial differential equation

$$u_t + (1 + u)u_x = 0. \tag{1.11}$$

This equation embodies the simplest type of nonlinearity (uu_x), and comparison with equation (1.3) might suggest that it is merely a case of replacing c by $1 + u$ in the solution. It turns out (by the method of characteristics) that this is correct! From equation (1.11) we see that

$$u = \text{constant on lines } \frac{dx}{dt} = 1 + u,$$

Fig. 1.1 A sketch of a wave packet, showing the wave and its envelope. The wave moves at the phase velocity, c, and the envelope at the group velocity, c_g.

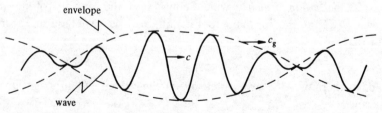

envelope

wave

and so the characteristic lines are $x = (1 + u)t + \text{constant}$. Thus the general solution is

$$u(x, t) = f\{x - (1 + u)t\}, \tag{1.12}$$

where f is an arbitrary function.

Now, given the initial wave profile, $u(x, 0) = f(x)$, it is a matter of solving equation (1.12) for u; this may be far from straightforward, even though the geometrical construction of the solution by characteristics is easy. In fact the solution of equation (1.12) (with $f > 0$, say, for some x) will generate a single-valued solution for u only for a finite time; thereafter the solution will be multi-valued (i.e. non-unique). The solution obtained by construction exhibits the non-uniqueness as a wave which has 'broken' (see Fig. 1.2). (Thus the solution must necessarily change its shape as it propagates.) This difficulty is usually overcome by the insertion of a jump (or *discontinuity*) which models a shock (again, see Fig. 1.2). Strictly, a discontinuous solution is not a proper solution of equation (1.11) but it may be allowable as a solution of the integral conservation equation from which (1.11) may have been derived.

Another complication arises with nonlinear equations: let us suppose that we have two solutions of equation (1.11), $u_1(x, t)$ and $u_2(x, t)$. We have already met the 'superposition principle' which says that, for linear equations, any linear combination of u_1 and u_2 is also a solution. However this is not true, in general, for nonlinear equations. It is easily verified that $u = u_1 + u_2$ does *not* satisfy equation (1.11). Thus solutions of nonlinear equations can not be superposed to form new solutions, although a related principle *is* available for certain nonlinear partial differential equations as we shall see.

It is clear that, by making suitable assumptions in a given physical problem, we might obtain an equation which is both nonlinear and

Fig. 1.2 The evolution of a nonlinear wave as time increases (a) $t = t_1$; (b) $t = t_2 > t_1$; (c) $t = t_3 > t_2$. The wave becomes vertical at one point at $t = t_2$, and thereafter the solution is triple-valued in a region. The solution can be made single-valued by the insertion of a discontinuity (and the smooth 'lobes' are then ignored).

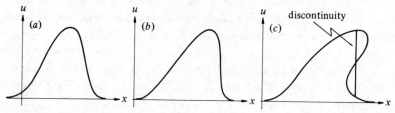

contains dispersive or dissipative terms (or both). So, for example, we might derive

$$u_t + (1 + u)u_x + u_{xxx} = 0, \tag{1.13}$$

or

$$u_t + (1 + u)u_x - u_{xx} = 0. \tag{1.14}$$

The first of these is the simplest equation embodying nonlinearity and dispersion; this, or one of its elementary variants, is known as the *Korteweg–de Vries* (or *KdV*) *equation*, of which we shall say much more later. The second one, equation (1.14), with nonlinearity and dissipation, is the *Burgers equation*. In fact the general solution of equation (1.14) has been known since 1906 (Forsyth, 1906), and it turns out that there are some pointers in the method of solution which are relevant to the solution of the KdV equation. (The properties of the Burgers equation will be left to the reader to explore in the exercises.) Our main concern will be with the method of solution – and the properties of – the KdV equation, and other related 'exactly integrable' equations. However, before we embark upon a more detailed discussion, the various alternative forms of the equation should be mentioned. We can transform equation (1.13) under

$$1 + u \rightarrow \alpha u, \qquad t \rightarrow \beta t, \qquad x \rightarrow \gamma x,$$

where α, β, γ are real (non-zero) constants, to yield

$$u_t + \frac{\alpha\beta}{\gamma} uu_x + \frac{\beta}{\gamma^3} u_{xxx} = 0.$$

This is a general form of the KdV equation, and a convenient choice, which we shall often use, is

$$u_t - 6uu_x + u_{xxx} = 0. \tag{1.15}$$

Some of these transformations of variables (as used above) belong to a *continuous group* or *Lie group*. As an example, consider the transformation, G_k, of the variables x, t and u into

$$X = kx, \qquad T = k^3 t, \qquad U = k^{-2}u,$$

for real $k \neq 0$. The application of successive transformations G_k and G_l is equivalent to the single transformation G_{lk}, thereby producing the multiplication law $G_l G_k = G_{lk}$. This law is commutative since $G_k G_l = G_{kl} = G_{lk}$. Furthermore, the associative law is also satisfied because $G_k(G_l G_m) = G_k G_{lm} = G_{klm} = G_{kl} G_m = (G_k G_l)G_m$. Clearly G_1 is the identity transformation: $G_1 G_k = G_{k1} = G_k$ for all k ($\neq 0$). If we form $G_{1/k}G_k = G_1$ and $G_k G_{1/k} = G_1$, we see that $G_{1/k}$ is both the left-hand and right-hand inverse of G_k. Therefore the elements of G_k for all real $k \neq 0$ form an *infinite group*. We call k the *parameter* of this continuous group. Now let us apply the transformation G_k to the KdV equation (1.15); it becomes

$$U_T - 6UU_X + U_{XXX} = 0,$$

i.e. it is *invariant* under the continuous group of transformations, G_k. This suggests that we seek invariant properties of the solutions. In particular, we anticipate the existence of *similarity solutions* which depend only on invariant combinations of the variables (see Q1.13 and section 2.6).

We have touched on ideas associated with waves in one spatial dimension, mainly because the KdV equation (and other equations we shall meet later) take this form. Of course, waves do occur in higher dimensions; in particular the classical wave equation can be written as

$$\frac{\partial^2 u}{\partial t^2} - c^2 \nabla^2 u = 0 \qquad (1.16)$$

where ∇^2 is the Laplace operator in the chosen coordinate system. It is clear that if we wished to examine more-complicated wave phenomena (with nonlinearity and dispersion), such as ring waves or waves crossing obliquely, then we must seek new equations. These might embody some of the character of both equations (1.15) and (1.16); in fact higher-dimensional KdV equations (and other integrable equations) do exist, but their discussion is beyond the scope of this text although one or two will be mentioned in the exercises.

1.2 The discovery of solitary waves

We have seen that the Korteweg–de Vries equation can be written down on the basis that both nonlinearity and dispersion might occur together. However, the KdV equation not only is of mathematical interest but also is of practical importance. To introduce this aspect, let us see how the solitary wave first appeared on the scientific scene. We shall then mention some of the analytical properties of this wave, and finally show that the KdV equation is indeed the relevant one for the solitary wave (and much more besides).

The solitary wave, so-called because it often occurs as this single entity and is localised, was first observed by J. Scott Russell on the Edinburgh–Glasgow canal in 1834; he called it the 'great wave of translation'. Russell reported his observations to the British Association in his 1844 'Report on Waves' in the following words:

I believe I shall best introduce the phenomenon by describing the circumstances of my own first acquaintance with it. I was observing the motion of a boat which was rapidly drawn along a narrow channel by a pair of horses, when the boat suddenly stopped – not so the mass of water in the channel which it had put in motion; it accumulated round the prow of the vessel in a state of violent agitation, then suddenly leaving it behind, rolled forward with great velocity, assuming the form of a large solitary elevation, a rounded, smooth and well-defined heap of water, which continued its course along the channel apparently without change of form or diminution of speed. I followed it on horseback, and overtook it still rolling on at a rate of some eight or nine miles an hour, preserving its original figure some thirty feet long and a foot to a foot and a half in height. Its height gradually diminished, and after a chase of one or two miles I lost it in the windings of the channel.

Russell also performed some laboratory experiments, generating solitary waves by dropping a weight at one end of a water channel (see Fig. 1.3). He was able to deduce empirically that the volume of water in the wave is equal to the volume of water displaced and, further, that the speed, c, of the solitary wave is obtained from

$$c^2 = g(h + a), \tag{1.17}$$

where a is the amplitude of the wave, h the undisturbed depth of water and g the acceleration of gravity (see Fig. 1.4). The solitary wave is therefore a *gravity wave*. We note immediately an important consequence of equation (1.17): higher waves travel faster. Fig. 1.3 and result (1.17) apply to waves of elevation; any attempt to generate a wave of depression results in a train of oscillatory waves, as Russell found in his own experiments.

To put Russell's formula (1.17) on a firmer footing, both Boussinesq (1871) and Lord Rayleigh (1876) assumed that a solitary wave has a length scale much greater than the depth of the water. They deduced, from the equations of motion for an inviscid incompressible fluid, Russell's formula for c. In fact they also showed that the wave profile $z = \zeta(x, t)$ is given by

$$\zeta(x, t) = a \operatorname{sech}^2 \{\beta(x - ct)\} \tag{1.18}$$

Fig. 1.3 Diagram of Scott Russell's experiment to generate a solitary wave.

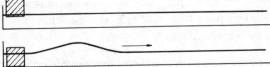

where $\beta^{-2} = 4h^2(h+a)/3a$ for any $a > 0$, although the sech^2 profile is strictly only correct if $a/h \ll 1$. These authors did not, however, write down a simple equation for $\zeta(x, t)$ which admits (1.18) as a solution. This final step was completed by Korteweg & de Vries in 1895. They showed that, provided ε and σ were small, then

$$\frac{\partial \zeta}{\partial t} = \frac{3}{2}\left(\frac{g}{h}\right)^{1/2}\left(\frac{2}{3}\varepsilon\frac{\partial \zeta}{\partial \chi} + \zeta\frac{\partial \zeta}{\partial \chi} + \frac{1}{3}\sigma\frac{\partial^3 \zeta}{\partial \chi^3}\right), \qquad (1.19)$$

where χ is a coordinate chosen to be moving (almost) with the wave. If we use the change of variables

$$\zeta = \zeta(X, t), \qquad X = \chi + \varepsilon\left(\frac{g}{h}\right)^{1/2}t$$

then equation (1.19) can be re-cast as the KdV equation

$$\zeta_t = \frac{3}{2}\left(\frac{g}{h}\right)^{1/2}\left(\zeta\zeta_X + \frac{1}{3}\sigma\zeta_{XXX}\right).$$

The parameter σ incorporates the surface tension, T, in the form $\sigma = \frac{1}{3}h^3 - Th/g\rho$, where ρ is the density of the liquid (and often $T \ll \frac{1}{3}g\rho h^2$); ε is an arbitrary parameter. We shall not reproduce the work of Korteweg & de Vries here, but it is instructive to see how the KdV equation arises from a set of fundamental governing equations. To this end we shall stay with water waves, but use the rather more satisfying technique of multiple-scale asymptotics.

* The governing equations of irrotational two-dimensional motion of an incompressible inviscid fluid, bounded above by a free surface and below by a rigid horizontal plane, are

$$\phi_{zz} + \delta^2\phi_{xx} = 0; \qquad \phi_z = 0 \quad \text{on } z = 0$$
$$\left.\begin{array}{l}\zeta + \phi_t + \frac{1}{2}\alpha(\delta^{-2}\phi_z^2 + \phi_x^2) = 0 \\ \phi_z = \delta^2(\zeta_t + \alpha\phi_x\zeta_x)\end{array}\right\} \text{on } z = 1 + \alpha\zeta, \qquad (1.20)$$

where ϕ is the velocity potential. The variables used here have already

Fig. 1.4 The parameters and variables used in the description of the solitary wave.

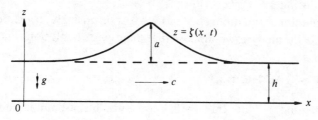

been non-dimensionalised by the use of the undisturbed depth, h, a typical horizontal length scale, l, and the typical speed $(gh)^{1/2}$. The surface is at $z = 1 + \alpha\zeta$ and on this surface we assume that the pressure is constant (so that in this simple theory the surface tension is ignored). The parameters appearing in equations (1.20) are given by

$$\alpha = a/h, \qquad \delta = h/l,$$

where a is a measure of the wave amplitude. The two boundary conditions on $z = 1 + \alpha\zeta$ describe the constancy of pressure at the surface, and the continuity of the vertical velocity component there.

We are interested in small-amplitude long waves i.e. in the limits as $\alpha \to 0$ and $\delta \to 0$. It turns out that one choice we could make – and which leads to the KdV equation for ζ – is to set $\delta^2 = O(\alpha)$ as $\alpha \to 0$. (This seems reasonable if we note the way in which α and δ appear in equations (1.20).) Clearly, however, this is rather special and we would hope that the solitary wave is a more enduring and general phenomenon than this would suggest. It is, as the following scaling for *arbitrary* δ demonstrates. Introduce

$$\xi = \frac{\alpha^{1/2}}{\delta}(x - t), \qquad \tau = \frac{\alpha^{3/2}}{\delta}t, \qquad \Phi = \frac{\alpha^{1/2}}{\delta}\phi; \qquad (1.21)$$

then equations (1.20) become

$$\begin{aligned} \Phi_{zz} + \alpha\Phi_{\xi\xi} = 0; \qquad \Phi_z = 0 \qquad & \text{on } z = 0 \\ \left.\begin{aligned} \zeta - \Phi_\xi + \alpha\Phi_\tau + \tfrac{1}{2}(\Phi_z^2 + \alpha\Phi_\xi^2) = 0 \\ \Phi_z = \alpha(-\zeta_\xi + \alpha\zeta_\tau + \alpha\Phi_\xi\zeta_\xi) \end{aligned}\right\} & \text{on } z = 1 + \alpha\zeta. \end{aligned} \qquad (1.22)$$

(Note that $\xi = O(1)$, $\phi = O(1)$ and $t = O(\alpha^{-1})$ if $\delta^2 = O(\alpha)$.) The choice of variables (1.21) means that equations (1.22) hold in a frame of reference which is moving with a speed of unity to the right, and then for large times (t) if $\tau = O(1)$ as $\alpha \to 0$ (for any fixed δ). In other words, scalings (1.21) describe a particular neighbourhood of (x, t)-space where we hope the KdV equation will be valid. The appearance of a speed of unity is by virtue of the non-dimensionalisation; this corresponds to a dimensional speed of $(gh)^{1/2}$ (cf. formula (1.17) for small a). Finally, the right-ward propagation is merely for convenience: we could equally well discuss left-ward motion by introducing $\xi = \alpha^{1/2}(x + t)/\delta$.

The solution of equations (1.22), as $\alpha \to 0$, is surprisingly straightforward. To initiate the analysis we suppose that there exists a solution which takes the form

$$\Phi \sim \sum_{n=0}^{\infty} \alpha^n \Phi_n(\xi, \tau, z), \qquad \zeta \sim \sum_{n=0}^{\infty} \alpha^n \zeta_n(\xi, \tau), \qquad \text{as } \alpha \to 0,$$

for fixed ξ and τ. (Note that $z \in [0, 1 + \alpha\zeta]$ which is a bounded domain if

ζ is bounded.) The leading order approximation now yields

$$\Phi_{0zz} = 0 \qquad \text{with} \qquad \Phi_{0z} = 0 \text{ on } z = 0,$$

and so $\Phi_0(\xi, \tau, z) \equiv \theta_0(\xi, \tau)$, say, an arbitrary function. Furthermore, the first surface boundary condition requires that $\zeta_0 = \Phi_{0\xi}$ on $z = 1$ (if we expand these conditions in a Taylor series about $z = 1$), and thus $\zeta_0 = \theta_{0\xi}$. If we continue the expansion for Φ then Laplace's equation, in conjunction with the bottom boundary condition, gives

$$\Phi \sim \theta_0 + \alpha(\theta_1 - \tfrac{1}{2}z^2\theta_{0\xi\xi}) + \alpha^2(\theta_2 - \tfrac{1}{2}z^2\theta_{1\xi\xi} + \tfrac{1}{24}z^4\theta_{0\xi\xi\xi\xi})$$

where $\theta_n = \theta_n(\xi, \tau)$, $n = 0, 1, 2$, are arbitrary functions. The surface boundary conditions now become

$$\zeta_0 + \alpha\zeta_1 - \{\theta_{0\xi} + \alpha(\theta_{1\xi} - \tfrac{1}{6}\theta_{0\xi\xi\xi})\} + \alpha\theta_{0\tau} + \tfrac{1}{2}\alpha\theta_{0\xi}^2 = O(\alpha^2)$$

and

$$-\alpha(1 + \alpha\zeta_0)\theta_{0\xi\xi} + \alpha^2(-\theta_{1\xi\xi} + \tfrac{1}{6}\theta_{0\xi\xi\xi\xi})$$
$$= \alpha(-\zeta_{0\xi} - \alpha\zeta_{1\xi} + \alpha\zeta_{0\tau} + \alpha\theta_{0\xi}\zeta_{0\xi}) + O(\alpha^3),$$

respectively. These two equations require that

$$\zeta_1 - \theta_{1\xi} + \tfrac{1}{6}\theta_{0\xi\xi} + \theta_{0\tau} + \tfrac{1}{2}\theta_{0\xi}^2 = 0$$

and

$$-\zeta_0\theta_{0\xi\xi} - \theta_{1\xi\xi} + \tfrac{1}{6}\theta_{0\xi\xi\xi\xi} = -\zeta_{1\xi} + \zeta_{0\tau} + \theta_{0\xi}\zeta_{0\xi},$$

where $\theta_{0\xi} = \zeta_0$. If we eliminate $\zeta_1 - \theta_{1\xi}$ then $\zeta_0(\xi, \tau)$ must satisfy

$$2\zeta_{0\tau} + 3\zeta_0\zeta_{0\xi} + \tfrac{1}{3}\zeta_{0\xi\xi\xi} = 0, \tag{1.23}$$

the KdV equation. (The interested reader might care to verify that the higher-order terms, ζ_n, satisfy an equation of the form

$$2\zeta_{n\tau} + \tfrac{3}{2}(\zeta_0\zeta_n)_\xi + \tfrac{1}{3}\zeta_{n\xi\xi\xi} = \mathscr{F}_{n-1}, \qquad n = 1, 2, \ldots,$$

where $\mathscr{F}_{n-1}$ denotes a function of $\zeta_0, \zeta_1, \ldots, \zeta_{n-1}, \zeta_{0\xi}, \ldots$.)

We have seen that the Korteweg–de Vries equation is indeed valid in an appropriate region of (x, t)-space, for small amplitude waves. However, we are left with one final connection to make: that between the KdV equation and the sech^2 profile. To demonstrate this, let us return to the equation as derived by Korteweg & de Vries themselves, equation (1.19). This has the advantage that it is written in physical variables and can therefore more readily be related to the work of Russell, Boussinesq and Rayleigh as expressed in equations (1.17) and (1.18). If the solution of equation (1.19) is stationary in the frame χ then $\zeta = \zeta(\chi)$ and so

$$\tfrac{2}{3}\varepsilon\zeta' + \zeta\zeta' + \tfrac{1}{3}\sigma\zeta''' = 0, \tag{1.24}$$

where the prime denotes the derivative with respect to χ. If we consider $\zeta \to 0$ as $|\chi| \to \infty$ (as is the case for the solitary wave) then equation (1.24) can be integrated twice to yield

$$2\varepsilon\zeta^2 + \zeta^3 + \sigma(\zeta')^2 = 0,$$

(the second integration requiring the integrating factor ζ'). This equation may be integrated once again (see §2.2), but it is more easily verified by direct substitution that

$$\zeta(\chi) = a \operatorname{sech}^2 (\beta\chi)$$

is a solution, provided

$$a = 4\sigma\beta^2 \qquad \text{and} \qquad \varepsilon = -2\sigma\beta^2.$$

The coordinate χ is defined (Korteweg & de Vries, 1895) as

$$\chi = x - (gh)^{1/2}\left(1 - \frac{\varepsilon}{h}\right)t$$

and so the solitary-wave solution becomes

$$\zeta(x, t) = a \operatorname{sech}^2\left[\frac{1}{2}\left(\frac{a}{\sigma}\right)^{1/2}\left\{x - (gh)^{1/2}\left(1 + \frac{1}{2}\frac{a}{h}\right)t\right\}\right]. \qquad (1.25)$$

This agrees with equations (1.17) and (1.18) if we neglect surface tension (so that $\sigma = \frac{1}{3}h^3$) and assume that $a/h \ll 1$, for then

$$c \sim (gh)^{1/2}\left(1 + \frac{1}{2}\frac{a}{h}\right) \qquad \text{and} \qquad \beta \sim \frac{1}{2}\left(\frac{3a}{h^3}\right)^{1/2}.$$

Thus Russell's solitary wave is a solution of the KdV equation.

In conclusion, let us make two observations concerning the solitary-wave result given in equation (1.25). With an amplitude of a, we see that the speed of the wave *relative to the speed of infinitesimal waves* (i.e. $(gh)^{1/2}$) is proportional to a. Also the 'width' of the wave (defined as the distance between the points of height $\frac{1}{2}a$, say) is inversely proportional to $a^{1/2}$. In other words, taller waves travel faster and are narrower. Finally, note how a appears in equation (1.25) and compare this with the way α appears in the scaled variables (1.21) that were used in our derivation of the KdV equation.

1.3 The discovery of soliton interactions

Hidden away in Russell's 'Report on Waves' (1844, see plate XLVII) is the diagram reproduced in Fig. 1.5, and the associated description. One interpretation of this result (with a little hindsight) is that an arbitrary initial profile (which in other words is not an *exact* solitary wave) will evolve into two (or more?) waves which then move apart and progressively approach individual solitary waves as $t \to \infty$. (Remember that our solitary wave is defined on $(-\infty, \infty)$.) This alone is rather surprising, but another remarkable property can also be observed. If we start with an initial profile like that given in Fig. 1.5, but with the taller wave somewhat to the *left* of the shorter, then the development is as depicted in Fig. 1.6. In this case the taller wave catches up, interacts with and then passes the shorter one. The taller one, therefore, appears to overtake the shorter one and continue on its way intact and undistorted. This, of course, is what we would expect if the two waves were to satisfy the linear superposition principle. But they certainly do not: this suggests that we have a special type of nonlinear process at work here. (In fact, the only indication that a linear interaction has *not* occurred is that the two waves are phase-shifted i.e. they are not in the positions after the interaction which would be anticipated if each were to move at a constant speed throughout the collision.)

The first hint that there was something unusual in the KdV equation

Fig. 1.5 A sketch of Scott Russell's 'compound' wave. This figure 'represents the genesis by a large low column of fluid of a compound or double wave of the first order, which immediately breaks down by spontaneous analysis into two, the greater moving faster and altogether leaving the smaller'. (Russell, 1844, p. 384)

Fig. 1.6 A sketch depicting the interaction of two 'solitons', for times (a) $t = t_1$; (b) $t = t_2 > t_1$; (c) $t = t_3 > t_2$; (d) $t = t_4 > t_3$.

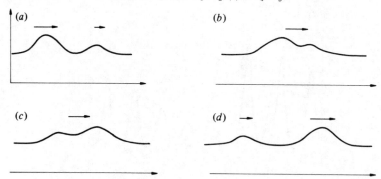

and solitary waves came in 1955. Fermi, Pasta & Ulam were working at Los Alamos on a numerical model of phonons in an anharmonic lattice, a model which turns out to be closely related to a discretisation of the KdV equation (Fermi, Pasta & Ulam, 1955). They observed that there was no equipartition of energy among the modes. Taking this up in 1965, Zabusky & Kruskal considered the initial-value problem for the equation

$$u_t + uu_x + \delta^2 u_{xxx} = 0, \tag{1.26}$$

with periodic boundary conditions (a more complicated problem than our infinite-domain solitary wave, but well-suited to numerical computation). They solved equation (1.26) with

$$u(x, 0) = \cos \pi x, \qquad 0 \leqslant x \leqslant 2,$$

and u, u_x, u_{xx} periodic on $[0, 2]$ for all t; they chose $\delta = 0.022$. A set of their results is shown in Fig. 1.7. After a short time the wave steepens and almost produces a shock, but the dispersive term ($\delta^2 u_{xxx}$) then becomes significant and some sort of local balance between nonlinearity and dispersion ensues. At later times the solution develops a train of eight well-defined waves, each like sech^2 functions, with the faster (taller) waves for ever catching-up and overtaking the slower (shorter) waves. (And there is another surprise: after a very long time, the initial profile – or something very close to it – reappears, a phenomenon requiring the topology of the torus for its explanation. This is an example of *recurrence*.)

At the heart of these observations is the discovery that these nonlinear waves can interact strongly and then continue thereafter almost as if there had been no interaction at all. This persistence of the wave led Zabusky & Kruskal to coin the name 'soliton' (after photon, proton, etc.), to emphasise

Fig. 1.7 The solution of the periodic boundary-value problem for the KdV equation (after Zabusky & Kruskal, 1965). Initial profile at $t = 0$ (dotted line); profile at $t = 1/\pi$ (broken line); profile at $t = 3.6/\pi$ (full line).

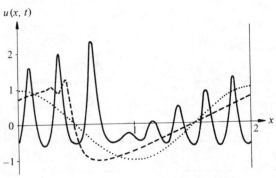

the particle-like character of these waves which seem to retain their identities in a collision. The discovery has led, in turn, to an intense study over the last twenty years. Many equations have now been found which possess similar properties, and diverse branches of pure and applied mathematics have had to be invoked to clarify many of the novel aspects that have appeared. We shall meet some of them in later chapters.

It is not easy to give a comprehensive and precise definition of a soliton. However, we shall associate the term with any solution of a nonlinear equation (or system) which (i) represents a wave of permanent form; (ii) is localised, so that it decays or approaches a constant at infinity; (iii) can interact strongly with other solitons and retain its identity. (There are more formal definitions – some of which concern discrete eigenvalues of a scattering problem – but these must wait until we have a more substantial mathematical framework.) In the context of the KdV equation, and other similar equations, it is usual to refer to the single-soliton solution as the *solitary wave*, but when more than one of them appear in a solution they are called *solitons*. Another way of expressing this is to say that the soliton becomes a solitary wave when it is infinitely separated from any other soliton. Also, we must mention the fact that for equations other than the KdV equation the solitary-wave solution may not be a sech^2 function; for example, we shall meet a sech function and also $\arctan(e^{\alpha x})$. Furthermore, some nonlinear systems have solitary waves but *not* solitons, whereas others (like the KdV equation) have solitary waves which *are* solitons.

1.4 Applications of the KdV equation

We have seen (in §1.1) that the KdV equation is the simplest equation we can envisage which incorporates both nonlinearity and dispersion. In fact it is easy to show that this equation should occur often in the description of real wave propagation. Consider a linear wave motion in one dimension with dispersion: we already know that the dispersion relation must take the form

$$\omega(k) = kc(k^2),$$

since only odd derivatives of u are allowed. (Our choice of dispersion, represented by a sum of derivative terms, will naturally produce a dispersion relation like $\omega = k\sum_{n=0}^{\infty} c_n k^{2n}$, but a more general functional dependence, $\omega = \omega(k)$, can arise which then has this form of expansion as $k \to 0$.) Now let us suppose that for infinitely long waves ($k \to 0$) there exists a non-zero speed of propagation, c_0, then

$$\frac{\omega}{k} \sim c_0 - \lambda k^2,$$

and usually long waves travel the fastest, and so $\lambda > 0$. This approximate dispersion relation is clearly obtained from the equation

$$u_t + c_0 u_x + \lambda u_{xxx} = 0.$$

Furthermore, if the medium in which the propagation is occurring is a classical continuum, then the time evolution will be given by the *material derivative* (or convective operator) $D/Dt = \partial/\partial t + u(\partial/\partial x)$. If these two effects are to balance then we shall obtain

$$u_t + c_0 u_x + \alpha(uu_x + \lambda u_{xxx}) = 0,$$

where α is a small parameter measuring the weak nonlinearity and long waves. Thus we have

$$u_\tau + uu_\xi + \lambda u_{\xi\xi\xi} = 0; \qquad \xi = x - c_0 t, \ \tau = \alpha t,$$

the KdV equation for small amplitude long waves, valid in an appropriate region of the (x, t)-plane (defined by $x - c_0 t = O(1)$, $t = O(\alpha^{-1})$, as $\alpha \to 0$.)

With these points in mind, we anticipate that the KdV equation will arise in a number of different contexts. We have already seen how the equation can be derived for the classical water-wave problem. A few of the many other applications include internal gravity waves in a stratified fluid, waves in a rotating atmosphere (Rossby inertial waves), ion-acoustic waves in a plasma and pressure waves in a liquid–gas bubble mixture. (The other equations which we shall meet later also have a wide application, and one of them – the nonlinear Schrödinger (or NLS) equation – is perhaps even more generally useful than the KdV equation.) In summary, we see that the KdV equation is a characteristic equation governing weakly nonlinear long waves whose phase speed attains a simple maximum for waves of infinite length.

Further reading

The following, referenced by sections, is intended to give some useful further reading.

1.1 For basic properties of linear and nonlinear waves, see Whitham (1974). For more information concerning group velocity, see Lighthill (1978). For the application of group transformations to differential equations, see Bluman & Cole (1974).

1.2 For another derivation of the KdV equation for water waves, see Kevorkian & Cole (1981); for other water-wave applications see Johnson (1973) for variable depth, Freeman & Johnson (1970) for waves on arbitrary shears and Johnson (1980) for a review of one- and two-dimensional KdV equations.

1.3 See the motion pictures of soliton interactions, particularly Zabusky, Kruskal & Deem (F1965) and Eilbeck (F1981). For a comparison of the KdV equation with water-wave experiments, see Hammack & Segur (1974).

1.4 A few of the many papers: internal gravity waves (Benney, 1966); Rossby waves (Benney, 1966; Redekopp & Weidman, 1978); ion-acoustic waves (Washimi & Taniuti, 1966); gas bubbles in a liquid (van Wijngaarden, 1968).

Exercises

Q1.1 Use the method of characteristics to derive d'Alembert's solution of the classical wave equation, (1.1).

Q1.2 Express d'Alembert's solution in terms of $u(x,0) = p(x)$ and $u_t(x,0) = q(x)$.

Q1.3 Find a relation between $p(x)$ and $q(x)$ in Q1.2 which produces a single wave-component travelling to the right.

Q1.4 Discuss the dispersion relation for the equation

$$u_t + u_x + u_{xxx} - u_{xx} = 0.$$

Q1.5 Compare the dispersion relations for the equations

$$u_t + u_x + u_{xxx} = 0$$

and

$$u_t + u_x - u_{xxt} = 0,$$

particularly in the limiting cases of long and short waves.

Q1.6 Obtain the solution of the equation

$$u_t + (1 + u)u_x = 0$$

with

$$u(x,0) = \begin{cases} u_0 x, & 0 \leqslant x \leqslant 1 \\ u_0(2 - x), & 1 \leqslant x \leqslant 2 \\ 0, & x \leqslant 0, x \geqslant 2, \end{cases}$$

where u_0 is a positive constant.

[Note that this initial profile is not differentiable at $x = 0, 1, 2$.]

Q1.7 By using the characteristics, sketch the solution of the problem in Q1.6 at various times.

Q1.8 Find the implicit solution of the equation

$$u_t + uu_x = 0$$

with $u(x,0) = \cos \pi x$. Show that u first has a point where u_x is infinite at $t = \pi^{-1}$. What form may the solution take if it is allowed to develop beyond $t = \pi^{-1}$?

Q1.9 *A general nonlinear wave.* Suppose that $u(x,t)$ satisfies the equation

$$u_t + c(u)u_x = 0, \qquad -\infty < x < \infty, \quad t > 0,$$

with $u(x,0) = f(x)$, where both f and c are differentiable. Use the method of characteristics to find the implicit solution, and hence deduce that u_x remains finite until $t = \min\limits_{-\infty < \lambda < \infty} (-[c'\{f(\lambda)\}f'(\lambda)]^{-1})$.

Q1.10 *Linear KdV dispersion.* If

$$u_t + u_{xxx} = 0$$

with $u(x,0) = f(x)$ and $u, u_x, u_{xx} \to 0$ as $|x| \to \infty$, use the Fourier transform to show that

$$u(x,t) = (3t)^{-1/3} \int_{-\infty}^{\infty} f(y)\mathrm{Ai}\left(\frac{x-y}{(3t)^{1/3}}\right)dy,$$

where $\mathrm{Ai}(z)$ is the Airy function of z.

Q1.11 *Solitary wave.* Obtain the solitary-wave solution of the equation

$$u_t - 6uu_x + u_{xxx} = 0,$$

for a wave of amplitude $-2\kappa^2$.

Q1.12 *Rational solitary wave.* Show that

$$u(x,t) = 6x\frac{(x^3 - 24t)}{(x^3 + 12t)^2}$$

is a solution of the KdV equation given in Q1.11. (Note that this solution is singular on $x^3 + 12t = 0$.)

[Hint: it might help to write $u = 6t^{-2/3}f(\eta)$, $\eta = xt^{-1/3}$ (see Q1.13), and then to observe that $f = -\frac{1}{3}(\log F)''$, $F(\eta) = \eta^3 + 12$.]

Q1.13 *Painlevé equation.* Show that the KdV equation

$$u_t - 6uu_x + u_{xxx} = 0$$

is invariant under the transformation $x \to kx$, $t \to k^3 t$, $u \to k^{-2}u$ ($k \neq 0$). Also verify that $t^{2/3}u$ and $xt^{-1/3}$ are invariant under the same transformation. Show that if $u(x,t) = -(3t)^{-2/3}F(\eta)$, where $\eta = x(3t)^{-1/3}$, then

$$F''' + (6F - \eta)F' - 2F = 0.$$

Hence, by setting $F = \lambda\,dV/d\eta - V^2$, $V = V(\eta)$, where λ is a constant to be determined, verify that after two integrations the equation for $V(\eta)$ can be written as

$$V'' - \eta V - 2V^3 = 0,$$

provided V decays exponentially as either $\eta \to +\infty$ or $\eta \to -\infty$.

[This equation for V is a Painlevé equation of the second kind; see Chap. 7 and Ince (1927).]

Q1.14 *KdV equation.* Suppose that the phase velocity of some linear wave is $c(k)$, where k is the wave number. Now weakly nonlinear waves can often be described by an equation of the form

$$u_t + uu_x + \int_{-\infty}^{\infty} K(x - \xi)u_\xi(\xi, t)\,d\xi = 0,$$

where the kernel K is determined from linear theory as the Fourier transform of c,

$$K(x) = \frac{1}{2\pi} \int_{-\infty}^{\infty} c(k)e^{ikx}\,dk.$$

For water waves it is well-known that $c^2 = (g/k)\tanh(kh)$ where g is the acceleration of gravity and h is the undisturbed depth of the water: use this information to justify the KdV equation for long waves.

*Q1.15 *Benjamin–Ono equation.* In Q1.14 take $c(k) = c_0(1 - \lambda|k|)$, where c_0 and λ are constants, and hence deduce that

$$u_t + (c_0 + u)u_x + \frac{\lambda c_0}{\pi} \fint_{-\infty}^{\infty} \frac{u_{\xi\xi}(\xi, t)}{\xi - x}\,d\xi = 0,$$

where $\fint$ denotes the Cauchy principal value, provided $u_\xi \to 0$ as $|\xi| \to \infty$.

[This is the Benjamin–Ono equation which arises in the study of internal waves: Davis & Acrivos, 1967; Benjamin, 1967; Ono, 1975.]

Elementary solutions of the Korteweg–de Vries equation

2.1 Travelling-wave solutions

A travelling wave of permanent form has already been met; this is the solitary-wave solution of the KdV equation itself. Such a wave is a special solution of the governing equation which does not change its shape and which propagates at constant speed. This wave may be localised or periodic. In the case of linear equations the profile is usually arbitrary, and is rarely of any special significance; a nonlinear equation, however, will normally determine a restricted class of profiles which often play an important rôle in the solution of the initial-value problem as $t \to \infty$. So, for example, the classical wave equation

$$u_{tt} - c^2 u_{xx} = 0$$

has the travelling-wave solutions $f(x - ct)$ and $g(x + ct)$, for *arbitrary f* and g (which together correspond to d'Alembert's solution). On the other hand the nonlinear equation

$$u_t + (1 + u)u_x = 0$$

has a travelling-wave solution $u(x, t) = f(x - ct)$ only if

$$(1 - c + f)f' = 0$$

and so $f = $ constant: a trivial (non-wave-like) solution. It is obvious that neither of these examples – as they stand – will teach us very much.

Let us restrict consideration to the more interesting area of nonlinear equations. The simplest such equation (mentioned above) does not have a travelling-wave solution at all, and this is to be expected from the general solution discussed in §1.1. The wave steepens and, if allowed, will 'break' and become multi-valued; at no stage is a steady profile possible. Similarly the effects of dispersion alone also produce a wave which forever changes its shape, but now in the opposite sense in that it causes the wave to spread out rather than to steepen. Perhaps these two effects may maintain a balance and thereby produce a wave of permanent form. Of course it is precisely this balance which gives rise to the solitary wave. (A similar balance can be struck between nonlinearity and dissipation; see Q2.1(i).)

As an example of the general method for seeking travelling-wave solutions, let us consider

$$u_t + (1 + u)u_x = v(u) \tag{2.1}$$

for some function $v(u)$. The required solution must take the form $u(x, t) = f(x - ct)$, where c is a constant which may play the rôle of a parameter (as in the KdV solitary wave, see §2.2) or it may be determined uniquely. Equation (2.1) now becomes

$$(1 - c + f)f' = v(f)$$

and so

$$\int \frac{(1 - c + f)}{v(f)} \, df = \xi, \qquad \text{where } \xi = x - ct.$$

Let us suppose that equation (2.1) is given with $v(u) = u(1 - u^2)$, and for simplicity we choose $c = 1$. (The problem for arbitrary c may be undertaken as an exercise.) Thus we have

$$\frac{1}{2} \int \left(\frac{1}{1 - f} + \frac{1}{f + 1} \right) df = \xi$$

which yields

$$\log \left| \frac{1 + f}{1 - f} \right| = A e^{2\xi} \qquad \text{or} \qquad f = \frac{A e^{2\xi} - 1}{A e^{2\xi} + 1},$$

where A is an arbitrary constant. This solution is more conveniently written as

$$u(x, t) = f(x - t) = \tanh(x - t - x_0),$$

where $A = \exp(-2x_0)$, and this describes a smoothed step propagating to the right.

2.2 Solitary waves

We now turn specifically to the KdV equation, and briefly discuss the solitary-wave solution mentioned in §1.2. It is convenient (particularly in view of the later work) to write the KdV equation in the standard form

$$u_t - 6uu_x + u_{xxx} = 0. \tag{2.2}$$

The travelling-wave solutions of this equation are $u(x, t) = f(\xi)$, where $\xi = x - ct$ and c is a constant. Thus equation (2.2) becomes

$$-cf' - 6ff' + f''' = 0,$$

which may be integrated once to yield

$$-cf - 3f^2 + f'' = A,$$

where A is an arbitrary constant. If we now use f' as an integrating factor we may integrate once more to give

$$\tfrac{1}{2}(f')^2 = f^3 + \tfrac{1}{2}cf^2 + Af + B, \tag{2.3}$$

where B is a second arbitrary constant. (We shall examine equation (2.3) for general A, B in the next section.) At this stage let us impose the boundary conditions $f, f', f'' \to 0$ as $\xi \to \pm \infty$ which describe the solitary wave. Thus A and B are both zero,

$$(f')^2 = f^2(2f + c) \tag{2.4}$$

(essentially as we quoted in §1.2), and we can see immediately that a real solution exists only if $(f')^2 \geqslant 0$ i.e. if $2f + c \geqslant 0$.

Equation (2.4) can be integrated as follows: first write

$$\int \frac{\mathrm{d}f}{f(2f + c)^{1/2}} = \pm \int \mathrm{d}\xi$$

and then use the substitution $f = -\tfrac{1}{2}c \operatorname{sech}^2 \theta \ (c \geqslant 0)$ to obtain

$$f(x - ct) = -\tfrac{1}{2}c \operatorname{sech}^2 \{\tfrac{1}{2}c^{1/2}(x - ct - x_0)\}, \tag{2.5}$$

where x_0 is an arbitrary constant of integration. Note that the choice $\pm$ is redundant since the solution is an even function, and also that the constant x_0 (a phase shift) plays a minor rôle: it merely denotes the position of the peak at $t = 0$. The solitary-wave solution (2.5) of equation (2.2) forms a one-parameter family (ignoring x_0), and in fact the solution exists for all $c \geqslant 0$ no matter how large or small the wave may be. (The solitary-wave solution of the *original* water-wave equations (1.20) exists only up to a maximum amplitude and its profile is only approximated by a sech^2 function.) The fact that $f \leqslant 0$ reflects our choice of KdV equation (2.2) with negative nonlinearity; we may recover the classical wave of elevation by transforming $u \to -u$.

2.3 General waves of permanent form

The *qualitative* nature of the solution $f(\xi)$ of equation (2.3), for arbitrary values of the constants c, A and B, can be determined by elementary analysis. The *quantitative* behaviour, however, requires the use of elliptic functions (see §2.4) or numerical computation.

It is clear that, for practical applications, we are interested only in real bounded solutions $f(\xi)$ of

$$\tfrac{1}{2}(f')^2 = f^3 + \tfrac{1}{2}cf^2 + Af + B \equiv F(f).$$

Thus we require (as before) $(f')^2 \geqslant 0$, and the form of $F(f)$ shows that f will vary monotonically until f' vanishes (i.e. $F(f)$ has at least one real zero). In other words, we can anticipate that the zeros of $F(f)$ are important.

A little thought shows that the zeros of $F(f)$, according to the values of c, A and B, must fall into one of the six categories depicted in Fig. 2.1. Since $(f')^2 \geqslant 0$, a real – but not necessarily bounded – solution will occur only in the intervals shaded in the figure. Further, to lay the foundations for our discussion, we need to consider the behaviour of f near a zero of $F(f)$ and clearly three cases can arise: $F(f)$ will have a simple, double or triple zero. Let $F(f_1) = 0$ and we then examine each case in turn.

(i) If f_1 is a *simple* zero then

$$(f')^2 = 2(f - f_1)F'(f_1) + O((f - f_1)^2),$$

as $f \to f_1$, which can be solved iteratively to yield

$$f = f_1 + \tfrac{1}{2}(\xi - \xi_1)^2 F'(f_1) + O((\xi - \xi_1)^3),$$

as $\xi \to \xi_1$, where $f(\xi_1) = f_1$. Thus f has a local minimum or maximum at $\xi = \xi_1$, as $F'(f_1)$ is positive or negative, respectively.

Fig. 2.1 Sketches of the graphs of $F(f)$ for the six different cases. Real solutions, $(f')^2 \geqslant 0$, will occur only in the shaded regions.

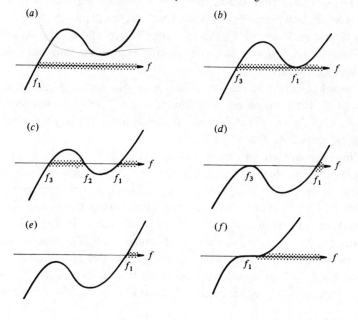

(ii) If f_1 is a *double* zero then

$$(f')^2 = (f - f_1)^2 F''(f_1) + O((f - f_1)^3),$$

as $f \to f_1$, and this is allowable only if $F''(f_1) > 0$ (see Fig. 2.1(*b*)). This time we obtain

$$f - f_1 \sim \alpha \exp[\pm \xi \{F''(f_1)\}^{1/2}] \qquad \text{as } \xi \to \mp \infty, \qquad (2.6)$$

if f is to be bounded; α is an arbitrary constant. Thus $f \to f_1$ as $\xi \to \mp \infty$ (signs vertically ordered throughout), and the solution can therefore have only one peak and the wave must extend from $-\infty$ to $+\infty$.

(iii) If f_1 is a *triple* zero then there is only one possibility, namely $f_1 = -c/6$ with $A = 3(c/6)^2$ and $B = (c/6)^3$: see Fig. 2.1(*f*). The complete solution for $f(\xi)$ is then easily obtained as

$$f(\xi) = -\frac{c}{6} + \frac{2}{(\xi - \beta)^2}, \qquad (2.7)$$

where β is an arbitrary constant. This solution is unbounded at $\xi = \beta$. (Note that there is always available the trivial solution $f \equiv f_1$ (for example, $\alpha = 0$ in expression (2.6) or $|\beta| \to \infty$ in (2.7)) but this is not a propagating-wave solution.) Thus, on ignoring (iii) which is not relevant, f' will either change sign across $f = f_1$ (see (i)) or $f' \to 0$ as $\xi \to \pm \infty$ (see (ii)).

Now consider the cases depicted in Figs 2.1(*a*), (*d*), (*e*) and the right-hand part of (*c*) (beyond $f = f_1$). If at some point $\xi = \xi_0$ (say) on the solution, the slope is such that $f' > 0$, then $F > 0$ for all $\xi > \xi_0$ and $f \to +\infty$ as $\xi \to +\infty$. If, however, $f'(\xi) < 0$ then f will decrease until it reaches $f = f_1$ (the largest real zero of F); this is a simple zero and so f' changes sign and once again $f \to +\infty$ as $\xi \to +\infty$. Hence, for these four cases, there is no bounded solution.

From Fig. 2.1(*b*) we see that F has a simple zero at f_3 and a double zero at f_1; the solution has a minimum at $f = f_3 (F'(f_3) > 0)$ and attains $f = f_1$ as $\xi \to \pm \infty$. This, of course, is therefore the solitary-wave solution with an amplitude $f_3 - f_1 (< 0)$.

Finally we are left with the left-hand part of curve (*c*) in the finite region where $(f')^2 \geqslant 0$. Here we have simple zeros at both f_2 and f_3; in fact there is a local maximum at f_2 (since $F'(f_2) < 0$) and a local minimum at f_3 (since $F'(f_3) > 0$). Thus f' will change sign at these points and since the behaviour near them is algebraic (not exponential as in (ii)), consecutive points $f = f_2, f = f_3$ will be a finite distance apart. The unique solution will be completely determined if f, and the sign of f', are given at any point $\xi = \xi_0$, which then fixes a point on the curve between f_2 and f_3. The solution will thereafter oscillate between f_2 and f_3, with a finite period.

This period can be expressed as

$$2 \int_{f_3}^{f_2} \frac{\mathrm{d}f}{f'} = 2 \int_{f_3}^{f_2} \frac{\mathrm{d}f}{\{2F(f)\}^{1/2}} \qquad (f_2 > f_3)$$

and the solution itself is given *implicitly* by

$$\xi = \xi_3 \pm \int_{f_3}^{f} \frac{\mathrm{d}f}{\{2F(f)\}^{1/2}} \qquad (2.8)$$

where $f(\xi_3) = f_3$ and the $\pm$ is according to $f' \gtrless 0$. Any further discussion of this solution requires the introduction of Jacobian elliptic functions, as used by Korteweg & de Vries themselves in 1895. In fact they named these new periodic solutions *cnoidal waves* (after 'cn', the relevant Jacobian elliptic function); we shall examine them more thoroughly in the next section, where two typical wave profiles are reproduced.

The methods used in this and the next section are applied to the KdV equation, but it should be noted that they may also be applied to many other nonlinear partial differential equations (or systems). The methods describe, qualitatively, the solutions $f(\xi)$ of any equation which can be written as

$$f'' = -\frac{\mathrm{d}V}{\mathrm{d}f},$$

for a given 'potential' function $V(f)$, where $f' = \mathrm{d}f/\mathrm{d}\xi$. We can at once obtain the integral

$$\tfrac{1}{2}(f')^2 = E - V(f) \equiv F(f),$$

for some constant of integration, E, an 'energy'. We shall see that there exist periodic solutions $f(\xi)$ if $F(f)$ is positive between two simple zeros of F, and there exist solitary-wave solutions if F is positive between a simple zero and a double zero. Also there exist monotone solutions such that $f(\xi) \to f_1$ as $\xi \to \pm \infty$ and $f(\xi) \to f_2$ as $\xi \to \mp \infty$, if F is positive between two double zeros, f_1 and f_2. This third case will arise later (in connection with the sine–Gordon equation); the solution is called a *kink* or *topological soliton* (see Q2.19).

Here is one final and quite general point before we leave travelling waves of permanent form. We have seen, particularly for the KdV equation, how such solutions can be found. However, there is no guarantee that such waves will exist in the practical – or even computational – sense. If these special solutions are unstable to small perturbations then any physical (or numerical) disturbance will eventually destroy them. So, strictly, our task is not complete until we have examined the stability of these waves to at least small (linear) disturbances. This aspect of the problem goes beyond the scope of this book, but such analyses have been undertaken. For example, Benjamin (1972) has shown that the shape of the solitary wave is stable (i.e. ultimately unchanged) to small distortions (although the effect upon the phase shift, i.e. position, of the wave over long times is still an open question). Indeed, the vast amount of numerical work over the years attests to the amazing stability of the solitary wave. Similarly, Drazin (1977) has shown that the cnoidal wave is also stable to small disturbances.

*2.4 Description in terms of elliptic functions

A more complete mathematical discussion of the implicit relation (2.8) requires a brief introduction to elliptic functions and integrals. We shall present here all the information necessary to enable us to describe the cnoidal wave.

First we define the integral

$$v = \int_0^\phi \frac{d\theta}{(1 - m \sin^2 \theta)^{1/2}}, \tag{2.9}$$

where we shall take m, the *parameter*, such that $0 \leqslant m \leqslant 1$. We may compare the integral (2.9) with the elementary integral

$$w = \int_0^\psi \frac{dt}{(1 - t^2)^{1/2}}, \tag{2.10}$$

where we use $t = \sin \theta$ so that $w = \arcsin \psi$ or $\sin w = \psi$, and so observe that integral (2.10) defines the inverse of the trigonometric function, sin. This led Jacobi (and also Abel) to define a new pair of inverse functions from (2.9)

$$\text{sn } v = \sin \phi, \qquad \text{cn } v = \cos \phi. \tag{2.11}$$

These are two of the *Jacobian elliptic functions*; they are usually written $\text{sn}(v|m)$, $\text{cn}(v|m)$ to denote the dependence on the parameter. (It is not unusual to work with the *modulus*, k, where $m = k^2$; however, m is slightly more advantageous in the present context.)

The two special cases $m = 0, 1$ enable integrals (2.9) and functions (2.11) to be reduced to elementary functions: if $m = 0$ then,

$$v = \phi \qquad \text{and so} \qquad \text{cn}(v|0) = \cos \phi = \cos v,$$

and if $m = 1$ the integral can be evaluated to yield

$$v = \text{arcsech} (\cos \phi) \qquad \text{and so} \qquad \text{cn}(v|1) = \text{sech } v.$$

It therefore follows that $\text{cn}(v|m)$ and $\text{sn}(v|m)$ are periodic functions for

$0 \leqslant m < 1$, but that periodicity is lost for $m = 1$. Now the period of cn and sn corresponds to the period 2π of cos and sin, and so the period of these elliptic functions can be written as

$$\int_0^{2\pi} \frac{d\theta}{(1 - m\sin^2 \theta)^{1/2}} = 4 \int_0^{\pi/2} \frac{d\theta}{(1 - m\sin^2 \theta)^{1/2}}.$$

This latter integral is the *complete elliptic integral of the first kind* ('complete' because it has fixed limits),

$$K(m) = \int_0^{\pi/2} \frac{d\theta}{(1 - m\sin^2 \theta)^{1/2}}. \tag{2.12}$$

It is immediately clear that $K(0) = \pi/2$, and it is also straightforward to show that $K(m)$ increases monotonically as m increases. In fact

$$K(m) \sim \tfrac{1}{2}\log\{16/(1 - m)\}$$

as $m \to 1^-$, and so $K(m) \to +\infty$ as $m \to 1^-$ (see Q2.7). Of course, this just demonstrates the infinite 'period' of the $\mathrm{cn}(v|1) = \mathrm{sech}\, v$ function.

Algebraic and differential relations between the Jacobian elliptic functions can also be obtained quite easily. For example

$$\mathrm{cn}^2 + \mathrm{sn}^2 = 1$$

(from the Pythagorean result for the trigonometric functions), and

$$\frac{d}{dv}\,\mathrm{cn} = \frac{d\phi}{dv}\frac{d}{d\phi}\,\mathrm{cn} = (1 - m\sin^2 \phi)^{1/2}\frac{d}{d\phi}\cos\phi,$$

which is usually expressed in terms of a third elliptic function $\mathrm{dn}(v|m) = (1 - m\sin^2 \phi)^{1/2}$, so that

$$\frac{d}{dv}\,\mathrm{cn} = -\,\mathrm{sn}\,\mathrm{dn}^{\dagger}$$

(and also we see that $\mathrm{dn}^2 + m\,\mathrm{sn}^2 = 1$). Note that, since the argument (v) and parameter (m) are the same throughout these identities, we have suppressed them altogether.

We may now use this knowledge to derive the solution of equation (2.3), for the travelling-wave, in the case depicted in Fig. 2.1(*c*). One possible method is to use the differential and algebraic relations directly to *verify* that there exists a solution of equation (2.3) in the form

$$f(\xi) = a + b\,\mathrm{cn}^2\{\alpha(\xi - \xi_3)|m\},$$

for certain a, b, α, m. This is left as an exercise for the reader. We shall

---

† It would be more accurate – but not relevant for us – to write $(\partial/\partial v)\,\mathrm{cn}$ since the derivative $(\partial/\partial m)\,\mathrm{cn}$ also exists.

adopt the following more systematic approach. The three distinct zeros of $F(f)$ are denoted by $f_3 < f_2 < f_1$ (see Fig. 2.1(c)), and so we can write

$$\xi = \xi_3 \pm \int_{f_3}^{f} \frac{dg}{\{2(g-f_1)(g-f_2)(g-f_3)\}^{1/2}} \qquad (2.13)$$

from equation (2.8). This is transformed into a standard elliptic integral by using the substitution

$$g = f_3 + (f_2 - f_3)\sin^2\theta$$

to give

$$\xi = \xi_3 \pm \{2/(f_1 - f_3)\}^{1/2} \int_{0}^{\phi} \frac{d\theta}{(1 - m\sin^2\theta)^{1/2}}$$

where $m = (f_2 - f_3)/(f_1 - f_3)$ and

$$f = f_3 + (f_2 - f_3)\sin^2\phi = f_2 - (f_2 - f_3)\cos^2\phi.$$

Thus we have

$$\mathrm{cn}[(\xi - \xi_3)\{(f_1 - f_3)/2\}^{1/2}|m] = \cos\phi,$$

where the $\pm$ is suppressed since cn is an even function, and so

$$f(\xi) = f_2 - (f_2 - f_3)\,\mathrm{cn}^2[(\xi - \xi_3)\{(f_1 - f_3)/2\}^{1/2}|m], \qquad (2.14)$$

the cnoidal-wave solution.

The shape of the cnoidal wave can now be obtained by direct computation – or from tables of the Jacobian elliptic functions – given values of f_1, f_2, f_3. One period of the wave is shown in Fig. 2.2 for two values of m, $0 < m < 1$. It is clear from solution (2.14) that the level $f = f_2$ describes the peak of the wave and $f = f_3$ the trough (since $0 \leqslant \mathrm{cn}^2 \leqslant 1$ and $f_3 < f_2$), and so $\frac{1}{2}(f_2 - f_3)$ could be regarded as the amplitude of the

Fig. 2.2 One period of the cnoidal wave, for $m = 0.6$, 0.9. The linear wave, $m = 0$, is included for comparison. All three waves have been normalised so that the amplitudes, and wave lengths, are the same. Note that we have plotted $-f$ so as to present the conventional wave of elevation as $m \to 1$.

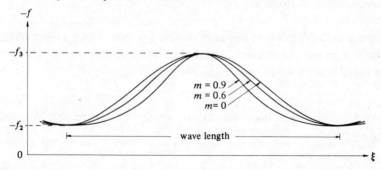

wave. The wave length can also be determined as

$$2K(m)\{2/(f_1 - f_3)\}^{1/2},$$

remembering that $cn^2(v|m)$ has a period $2K(m)$, not $4K(m)$. Finally the shape of the wave is governed by the value of the parameter m, as Fig. 2.2 demonstrates. Of course we have all along been describing a travelling-wave solution with $\xi = x - ct$, and by comparing equations (2.3) and (2.13) we see that the speed of propagation is $c = -2(f_1 + f_2 + f_3)$. Thus solution (2.14) represents a strongly nonlinear wave: the speed, shape and wave length (or period) all depend on the *amplitude* of the wave in a quite complicated way. Any particular cnoidal wave will be completely determined if the peak, trough and wave length are prescribed, for these will fix f_1, f_2 and f_3. (Note that for the water-wave problem, f_2 and f_3 are measured relative to the undisturbed level of the water: see the derivation of the KdV equation in §1.2.) Cnoidal waves can sometimes be observed in rivers, although often with slowly varying amplitude and period as in the case of the train of waves behind a weak bore (the so-called *undular bore*).

Finally, there are two points of some mathematical interest which should not be ignored. We have seen that the dependence on the amplitude is quite involved, but we anticipate some significant and instructive simplification if we let the amplitude tend to zero (i.e. the 'linear' limit, which corresponds to $m \to 0$). On the other hand, the 'most nonlinear' limit of $m \to 1$ should also be informative: as $m \to 1$ we should recover the solitary-wave solution. These two limits are now examined in a little detail.

*2.5 Limiting behaviours of the cnoidal wave

First we let the amplitude tend to zero: let $\frac{1}{2}(f_2 - f_3) = a$, and then $m = 2a/(f_1 - f_3)$ so that $m \to 0$ as $a \to 0$. Now, we have

$$cn(v|m) \to \cos v \qquad \text{as } m \to 0,$$

and therefore

$$f \sim f_2 - 2a\cos^2\left[(\xi - \xi_3)\{(f_1 - f_3)/2\}^{1/2}\right] \qquad (2.15)$$

as $a \to 0$. The speed of propagation also takes a limiting form,

$$c \to -2(f_1 + 2f_2) \qquad \text{as } a \to 0.$$

The solution (2.15) can be expressed more conveniently if we introduce

$$k = \{2(f_1 - f_3)\}^{1/2} \qquad \text{and} \qquad \hat{f}_2 = f_2 - a;$$

we then obtain

$$f = \hat{f}_2 - a \cos \{k(x - ct - \xi_3)\} + O(a^2) \qquad \text{as } a \to 0,$$

which is a linear wave of amplitude a oscillating about the mean level $f = \hat{f}_2$. Furthermore, we can see that

$$\begin{aligned}
\omega = kc &= -2k(f_1 + 2\hat{f}_2) \\
&= -k(k^2 + 6\hat{f}_2) + O(a) \\
&\to -6k\hat{f}_2 - k^3 \qquad \text{as } a \to 0,
\end{aligned}$$

which is just the dispersion relation for the equation

$$u_t - 6\hat{f}_2 u_x + u_{xxx} = 0.$$

This, in turn, can be obtained by linearising the original KdV equation (2.2) about $u = \hat{f}_2$ (i.e. set $u \to \hat{f}_2 + u, |u| \ll 1$). In other words, the limiting behaviour of the cnoidal wave (as $m \to 0$) generates a linear wave with the correct dispersion relation.

The solitary-wave limit requires the two simple zeros at $f = f_1, f_2$ to coalesce to form a double zero (i.e. Fig. 2.1(c) → (b)). To accomplish this we let $f_2 \to f_1^-$ (at fixed f_3), and this implies that $m \to 1^-$. Again, recalling that

$$\text{cn}(v|m) \to \text{sech } v \qquad \text{as } m \to 1^-$$

we obtain

$$f \to f_1 - (f_1 - f_3) \text{sech}^2 [(\xi - \xi_3)\{(f_1 - f_3)/2\}^{1/2}]$$

where $\xi = x - ct$ with $c \to -2(2f_1 + f_3)$ as $f_2 \to f_1$. Now we set $f_1 - f_3 = \frac{1}{2}a$, the amplitude of the wave, and hence

$$f \to f_1 - \frac{1}{2}a \text{ sech}^2 \{\frac{1}{2}a^{1/2}(X - at - \xi_3)\} \tag{2.16}$$

where $X = x + 6f_1 t$ is a coordinate moving at a speed consistent with the ambient level $f = f_1$. (Of course, we could always choose $f_1 = 0$: this merely readjusts the undisturbed level below the solitary wave.) Solution (2.16) agrees with the solution (2.5) (where for c read a), obtained by transforming the original KdV equation (2.2) under

$$u \to f_1 + u, \qquad (x, t) \to (x + 6f_1 t, t).$$

Thus the limit $m \to 1^-$ recovers the classical KdV solitary wave, as we might expect.

2.6 Other solutions of the KdV equation

The solutions of the KdV equation described in the foregoing sections do not exhaust the possibilities. Other fairly simple types of solution also exist, some of which have already been touched on in the exercises at the end of Chap. 1. The two particular alternatives of interest to us here are the *similarity solutions* and the *rational solutions*.

Similarity solutions are encountered in elementary studies of partial differential equations, being a standard procedure for reducing them to ordinary differential equations. Thus, for example, if $u(x, t)$ satisfies a given equation then we may seek a solution of the form

$$u(x, t) = t^m f(\eta) \qquad \text{where } \eta = x t^n$$

and m, n are to be chosen so that $f(\eta)$ satisfies an ordinary differential equation. Often both m and n are uniquely determined, but sometimes these may involve a free parameter so that we can set $m = 0$, for example, and thereby generate what is usually the simplest solution in this class. For example, the nonlinear equation

$$u_t + u u_x = 0$$

has the solution $u(x, t) = t^m f(x t^n)$ if $m + n = -1$ and

$$mf - (1 + m)\eta f' + f f' = 0.$$

The choice $m = 0$ then yields $f' = 0$ so that $f = \text{constant}$, or $f = \eta$. Thus $u(x, t) = x/t$ is a similarity solution of the nonlinear wave equation.

Similarly, the model dispersive equation

$$u_t + u_{xxx} = 0$$

has the solution $u(x, t) = f(x t^{-1/3})$, on having again chosen $m = 0$, where

$$-\tfrac{1}{3}\eta f' + f''' = 0$$

which can be solved in terms of Airy functions (cf. Q1.10). If we combine these two examples in the form

$$u_t - 6 u u_x + u_{xxx} = 0$$

we have the KdV equation which can be solved by substituting

$$u(x, t) = -(3t)^{-2/3} f(\eta), \qquad \eta = x/(3t)^{1/3}.$$

(See Q1.13. Note that here both m and n are fixed, and that they can be determined by requiring that f and η are invariant under a group of transformations which leaves the KdV equation invariant. The numerical factors are purely for convenience.) The equation for $f(\eta)$ is then

$$f''' + (6f - \eta)f' - 2f = 0$$

which can be reduced to a Painlevé equation (again see Q1.13) with a solution describing a wave profile which decays as $\eta \to +\infty$, and oscillates as $\eta \to -\infty$. The appearance of a Painlevé equation (for which each movable singularity is a pole: see Chap. 7 and Ince, 1927) is not by chance. It is currently thought that there is a direct correspondence between the occurrence of a Painlevé equation for a given partial differential equation, and the existence of an 'inverse scattering transform' (and therefore soliton solutions) for that equation.

Finally we take a brief look at rational solutions, that is, solutions which are rational functions of the independent variables. These are usually more difficult to find directly, unless we have some idea of their form (or can at least assume that they take some appropriate form). As it happens we have already met a rational solution: $u(x, t) = x/t$ is a rational solution of

$$u_t + uu_x = 0,$$

as well as being a similarity solution. (We could seek this one by assuming that u is separable; clearly a very special choice.) The KdV equation

$$u_t - 6uu_x + u_{xxx} = 0$$

also has a simple rational solution. Let us assume that $u = u(x)$ only, and that $u, u', u'' \to 0$ as $|x| \to \infty$, then

$$-6uu' + u''' = 0$$

which can be integrated twice to yield

$$(u')^2 = 2u^3.$$

This can be solved immediately to give

$$u(x, t) = 2/x^2,$$

which is chosen to be singular at $x = 0$. This is essentially the solution (2.7) for the case of a triple zero. The next solution in a *hierarchy* of 'rational solitons' for the KdV equation is

$$u(x, t) = 6x(x^3 - 24t)/(x^3 + 12t)^2$$

(see Q1.12). All these KdV rational solutions turn out to be singular, but for some of the other 'exactly integrable' equations this is not the case – they prove to be practical and useful solutions. There is, nevertheless, a connecting theme: all the rational solutions can be obtained by examining corresponding solitary-wave or soliton solutions in an appropriate limit. This idea goes a bit beyond the aims of this text, and the method is left for the reader to explore in Q2.17.

Further reading

2.1 For travelling waves, etc., see Whitham (1974).

2.2 For more on solitary waves see Whitham (1974), and Stoker (1957) specifically for their relevance to water waves.

2.3 ⎱ For properties of elliptic functions and integrals, see Abramowitz & Stegun (1964,
2.4 ⎰ Chaps. 16 & 17). For more detailed information, and lists of integrals, see Byrd &
2.5 ⎰ Friedman (1971). The historical development of the elliptic functions also makes
   interesting reading: see Kline (1972) and Bell (1937). For the shape of cnoidal waves,
   see Wiegel (1960).

2.6 For a detailed discussion of both the rôle of the Painlevé transcendents, and rational
   solitons, see Ablowitz & Segur (1981, Chap. 3). The Painlevé equations are discussed
   comprehensively in Ince (1927).

Exercises

Q2.1 Find the travelling-wave solutions, in the form $u(x, t) = f(x - ct)$, for each
   of the following equations:

   (i) *Burgers equation*

   $$u_t + uu_x = u_{xx},$$

   with $u \to 0$ as $x \to + \infty$ and $u \to u_0 (> 0)$ as $x \to - \infty$;

   (ii) *Modified KdV equation or mKdV equation*

   $$u_t + 6u^2 u_x + u_{xxx} = 0, \quad \text{with } u, u_x, u_{xx} \to 0 \text{ as } |x| \to \infty;$$

   (iii) *A generalised KdV equation*

   $$u_t + (n + 1)(n + 2)u^n u_x + u_{xxx} = 0,$$

   where $n = 1, 2, \ldots$, with $u, u_x, u_{xx} \to 0$ as $|x| \to \infty$;

   (iv) *An elastic-medium equation*

   $$u_{tt} = u_{xx} + u_x u_{xx} + u_{xxxx}, \quad \text{with } u_x, u_{xx}, u_{xxx} \to 0 \text{ as } |x| \to \infty.$$

Q2.2 In Q2.1 (iii), with the sign of the nonlinear term now negative, show that
   solitary-wave solutions exist only if n is odd.
   (Zabusky, 1967)

*Q2.3 *Nonlinear Schrödinger equation.* Consider the equation

$$iu_t + u_{xx} + u|u|^2 = 0,$$

and seek a travelling-wave solution in the form

$$u = re^{i(\theta + nt)},$$

where $r(x - ct)$ and $\theta(x - ct)$ are real functions, and c and n are real
constants. Show that

$$\theta' = \tfrac{1}{2}(c + A/S) \quad \text{and} \quad (S')^2 = -2F(S),$$

with $S = r^2$ and $F(S) = S^3 - 2(n - \tfrac{1}{4}c^2)S^2 + BS + A$, where A and B are
arbitrary (real) constants of integration.

   Examine the nature of the zeros of the cubic F, and hence (briefly)
discuss the occurrence and properties of periodic solutions for u. In
particular show that there exist solitary-wave solutions of the form

$$u(x, t) = ae^{i\{\frac{1}{2}c(x - ct) + nt\}} \operatorname{sech} \{a(x - ct)/\sqrt{2}\}$$

for all $a^2 = 2(n - \tfrac{1}{4}c^2) > 0$.

[The nonlinear Schrödinger equation is often abbreviated as the *NLS equation*, and sometimes called the *cubic Schrödinger equation*.]

*Q2.4 *Another nonlinear Schrödinger equation.* Consider the equation

$$iu_t + u_{xx} - u|u|^2 = 0,$$

and follow the same procedure as given in Q2.3. Hence show that there exist solitary-wave solutions with

$$r^2(\xi) = m - 2\kappa^2 \operatorname{sech}^2 \kappa\xi \qquad \text{and} \qquad c\tan\{\theta(\xi)\} = -2\kappa \tanh \kappa\xi,$$

for all c, where $\xi = x - ct$, $n = -m$ and $\kappa = \frac{1}{2}(2m - c^2)^{1/2}$ with $m > \frac{1}{2}c^2$.
(Zakharov & Shabat, 1973)

Q2.5 *Fisher's equation.* Seek travelling-wave solutions of the equation

$$u_t = u_{xx} + \alpha^2 u(1 - u),$$

where $\alpha > 0$ is a constant, in the form $u(x,t) = f(x - ct)$. Investigate the nature of these solutions by examining the phase-plane (f, f') and show, in particular, that there exist solutions such that $f \to 0$ as $x \to -\infty$, $f \to 1$ as $x \to +\infty$, for all $c \leqslant -2\alpha$.

[This equation originally arose in a theory of gene selection in a species: see Fisher (1937). It also arises in the theory of combustion, and chemical kinetics; Kolmogorov, Petrovsky & Piscounov, 1937; Aris, 1975; Fife, 1979.]

Q2.6 *A piecewise linear equation.* Seek travelling-wave solutions of the equation

$$u_t + u_x \operatorname{sgn} u + u_{xxx} = 0$$

in the form $u(x,t) = f(x - ct)$. For what values of the constant c may there exist solitary waves such that $f', f'' \to 0$ as $x \to \pm \infty$ but $f \to \text{constant} (\neq 0)$? Write down the solutions f, explicitly, for all the solitary waves. Do periodic-wave solutions for f exist?

Q2.7 *Elliptic functions and integrals.* Show that

(i) $$K(m) = \frac{\pi}{2} F(\tfrac{1}{2}, \tfrac{1}{2}; 1; m),$$

where $F(a, b; c; z)$ is the hypergeometric function;

(ii) $$K(m) \sim \tfrac{1}{2} \log\{16/(1 - m)\} \qquad \text{as } m \to 1^-$$

[hint: write $d\theta = (1 - m^{1/2} \sin\theta + m^{1/2} \sin\theta) d\theta$];

(iii) $$\frac{d}{du}(\operatorname{sn} u) = \operatorname{cn} u \operatorname{dn} u; \qquad \frac{d}{du}(\operatorname{dn} u) = -m \operatorname{sn} u \operatorname{cn} u.$$

*Q2.8 *Burgers equation.* Determine the values of m and n so that $u(x,t) = t^m f(xt^n)$ is a solution of the equation

$$u_t + uu_x = u_{xx},$$

and write down the equation satisfied by f. Hence obtain the solution for which $f \to 0$ as $x \to \infty$, and $f(0) = -2/\pi^{1/2}$.

Q2.9 *Modified KdV equation.* Show that the equation

$$u_t + 6u^2 u_x + u_{xxx} = 0$$

is invariant under the transformation $x \to \lambda x$, $t \to \lambda^3 t$, $u \to \lambda^{-1} u$ ($\lambda \neq 0$). Hence introduce $u(x, t) = t^{-1/3} f(xt^{-1/3})$ and show that f satisfies

$$f'' - \tfrac{1}{3}\eta f + 2f^3 = 0, \qquad \eta = xt^{-1/3},$$

provided $f \to 0$ sufficiently rapidly at infinity.
[Cf. Q1.13 and equation (7.5); Zakharov & Shabat, 1973.]

Q2.10 *Concentric KdV equation.* Seek a similarity solution of the equation

$$2u_t + t^{-1} u - 3uu_x + \tfrac{1}{3} u_{xxx} = 0$$

in the form $u(x, t) = -\tfrac{1}{3}(2/t^2)^{1/3} f\{x(2t)^{-1/3}\}$. If $f \to 0$ sufficiently rapidly as $x \to +\infty$, show that

$$v'' - \eta v + v^3 = 0 \qquad \text{where } \eta = x(2t)^{-1/3} \text{ and } f = v^2.$$

[In fact v is a Painlevé transcendent, cf. §7.1.1; Maxon & Viecelli, 1974; Miles, 1978a; Johnson, 1980. The equation for u is sometimes called the *cylindrical KdV equation.*]

Q2.11 *Nonlinear Schrödinger equation.* Show that the equation

$$iu_t + u_{xx} + vu|u|^2 = 0,$$

where v is a real constant, is invariant under each of the group transformations:

(i) $t \to t + \lambda$, $x \to x$, $u \to u$;
(ii) $t \to t$, $x \to x + \lambda$, $u \to u$;
(iii) $t \to \lambda^2 t$, $x \to \lambda x$, $u \to \lambda^{-1} u$ ($\lambda \neq 0$).

Hence use property (iii) to suggest a similarity solution of the form $u(x, t) = t^m f(xt^n)$, for suitable m and n, and obtain the equation for f.

Q2.12 *Modified KdV equation.* Show that the equation

$$u_t + 6u^2 u_x + u_{xxx} = 0$$

has the rational solution

$$u(x, t) = c - 4c/\{4c^2(x - 6c^2 t)^2 + 1\}$$

for any real constant c. Is this solution singular?
(Zabusky, 1967)

Q2.13 *Benjamin–Ono equation.* Given that the Hilbert transform, $\mathscr{H}(u)$, is defined by

$$\mathscr{H}(u) = \frac{1}{\pi} \int_{-\infty}^{\infty} \frac{u(y, t)}{y - x} \, dy,$$

show that $\mathscr{H}\{a/(x^2 + a^2)\} = -x/(x^2 + a^2)$, $a > 0$. Hence, or otherwise, verify that

$$u(x, t) = ba^2/\{(x - ct)^2 + a^2\}$$

is a rational solution of the equation

$$u_t + uu_x + \mathcal{H}(u_{xx}) = 0$$

provided that a and b are related to c. What are these relations?

(Davis & Acrivos, 1967; Benjamin, 1967; Ono, 1975)

Q2.14 *Nonlinear Schrödinger equation.* Verify that the equation

$$iu_t + u_{xx} + u|u|^2 = 0$$

has the rational-cum-oscillatory solution

$$u(x,t) = e^{it}\{1 - 4(1 + 2it)/(1 + 2x^2 + 4t^2)\}.$$

(Peregrine, 1983)

*Q2.15 *Breather.* Transform the modified KdV equation

$$u_t + 6u^2 u_x + u_{xxx} = 0$$

into

$$(1 + \phi^2)(\phi_t + \phi_{xxx}) + 6\phi_x(\phi_x^2 - \phi\phi_{xx}) = 0$$

where $u = v_x$, $\phi = \tan(\tfrac{1}{2}v)$ and $v \to 0$ as $|x| \to \infty$.

Hence, or otherwise, verify that

$$u(x,t) = -2\frac{\partial}{\partial x}\arctan\left(\frac{l\sin(kx + mt + a)}{k\cosh(lx + nt + b)}\right)$$

is a solution if

$$m = k(k^2 - 3l^2) \qquad \text{and} \qquad n = l(3k^2 - l^2),$$

with a and b arbitrary constants.

[This oscillatory-pulse soliton is called a *breather* or *bion*.]

Q2.16 *The Ma solitary wave.* Verify that the nonlinear Schrödinger equation

$$iu_t + u_{xx} + u|u|^2 = 0$$

has solutions of the form

$$u(x,t) = a\exp(ia^2 t)\{1 + 2m(m\cos\theta + in\sin\theta)/f\},$$

for all real a and m, where $n^2 = 1 + m^2, \theta = 2mna^2 t$ and $f(x,t) = n\cosh(ma\sqrt{2}x) + \cos\theta$.

(Ma, 1979; Peregrine, 1983)

Q2.17 *A singular solution.* Show that

$$u(x,t) = 2k^2 \operatorname{cosech}^2\{k(x - 4k^2 t)\}$$

is a singular solution of the KdV equation

$$u_t - 6uu_x + u_{xxx} = 0.$$

Now let $k \to 0$ (at fixed x, t) and hence obtain the rational soliton $u(x,t) = 2/x^2$.

Further, show that the singular solution can be obtained from the classical solitary-wave solution,

$$u(x,t) = -2k^2 \operatorname{sech}^2 \{k(x - 4k^2 t) - x_0\}$$

by setting $\exp(2x_0) = -1$.

(Ablowitz & Segur, 1981)

Q2.18 *The Gardner equation.* You are given a mixed KdV–mKdV equation, namely

$$u_t - 6uu_x + u_{xxx} = 12\delta u^2 u_x$$

for some constant δ. To seek waves of permanent form, assume that $u(x,t) = f(\xi)$, where $\xi = x - ct$, and deduce that

$$\xi = \int \frac{df}{\{2F(f)\}^{1/2}},$$

where $F(f) = \delta f^4 + f^3 + \frac{1}{2}cf^2 + Af + B$ for arbitrary constants A, B and c. By a geometrical argument, or otherwise, show that periodic solutions may exist for all δ; that if $\delta > 0$ then either a solitary wave or a kink (i.e. a topological soliton) may exist; that if $\delta < 0$ then a solitary wave may exist but not a kink.

[Cf. equation (5.11).]

Q2.19 *Sine–Gordon equation.* Verify that

$$\phi(x,t) = 4 \arctan [C \exp \{(x - \lambda t)/(1 - \lambda^2)^{1/2}\}]$$

is a solution of the sine–Gordon equation

$$\phi_{xx} - \phi_{tt} = \sin \phi$$

(written in laboratory coordinates), for arbitrary real constants C and λ ($|\lambda| < 1$).

[A solution for which ϕ increases by 2π is often called a *kink*, and one which decreases by 2π an *antikink*.]

Q2.20 *Breather for the sine–Gordon equation.* Verify that

$$\phi(x,t) = 4 \arctan \left(\frac{(1 - \lambda^2)^{1/2}}{\lambda} \frac{\sin \{\lambda(t - t_0)\}}{\cosh \{(1 - \lambda^2)^{1/2}(x - x_0)\}} \right)$$

is a solution of the sine–Gordon equation (see Q2.19), where x_0, t_0 and λ ($0 < |\lambda| < 1$) are arbitrary real constants.

[This solution does not propagate, and is therefore a *stationary soliton*; see §8.1.]

Q2.21 *A moving breather.* Show that if $x' = \gamma(x - vt)$ and $t' = \gamma(t - vx)$ where $\gamma = 1/(1 - v^2)^{1/2}$ and $-1 < v < 1$, then

$$\frac{\partial^2 \phi}{\partial x^2} - \frac{\partial^2 \phi}{\partial t^2} = \frac{\partial^2 \phi}{\partial x'^2} - \frac{\partial^2 \phi}{\partial t'^2}.$$

Deduce that the sine–Gordon equation is invariant under this transformation.

Hence, or otherwise, show that the sine–Gordon equation has the solution

$$\phi(x,t) = 4 \arctan \left(\frac{(1-\lambda^2)^{1/2}}{\lambda} \frac{\sin\{\gamma\lambda(t - vx - t_0)\}}{\cosh\{\gamma(1-\lambda^2)^{1/2}(x - vt - x_0)\}} \right)$$

for arbitrary real constants x_0, t_0, v and λ (with $-1 < v$, $\lambda < 1$ and $\lambda \neq 0$).

[The transformation used here is the special *Lorentz transformation* with speed of light unity. This solution is a moving breather; cf. Q2.20 and see §8.1.]

Q2.22 *A kink–antikink solution.* Verify that

$$\phi(x,t) = 4 \arctan \left(\frac{\lambda \cosh\{x/(1-\lambda^2)^{1/2}\}}{\sinh\{\lambda t/(1-\lambda^2)^{1/2}\}} \right),$$

where $0 < |\lambda| < 1$, is an exact solution of the sine–Gordon equation (see Q2.19). Interpret the solution as the interaction of a kink and an antikink by examining the asymptotic behaviour of ϕ as $t \to \pm\infty$.

[A sketch of the solutions is useful; note that ϕ is an even function of x, but an odd function of t, and that it is instantaneously a constant at $t = 0$: see §8.1.]

Q2.23 *Two-dimensional Korteweg–de Vries equation.* Show that the 2D KdV equation

$$(u_t - 6uu_x + u_{xxx})_x + 3u_{yy} = 0$$

(sometimes also called the *Kadomtsev–Petviashvili equation*) has the solitary-wave solution

$$u(x,t) = -\tfrac{1}{2}k^2 \operatorname{sech}^2 \{\tfrac{1}{2}(kx + ly - \omega t)\}$$

where $\omega = k^3 + 3l^2/k$.

(Kadomtsev & Petviashvili, 1970; Freeman, 1980.)

Q2.24 *Another 2D KdV equation.* Show that the equation

$$(u_t - 6uu_x + u_{xxx})_x - 3u_{yy} = 0$$

has the rational solution

$$u(x,t) = -4 \frac{(p^2 y^2 - X^2 + p^{-2})}{(p^2 y^2 + X^2 + p^{-2})^2},$$

where $X = x + p^{-1} - 3p^2 t$ and p is a real constant.

[Note that $y \to iy$ gives the 2D KdV equation of Q2.23, and that this rational solution is therefore singular for that equation; see Kadomtsev & Petviashvili (1970), Freeman (1980, §8.1).]

The scattering and inverse scattering problems

3.1 Preamble

The first and most important task that we shall undertake in this text is the solution of the general initial-value problem for the KdV equation. (This will be followed in Chap. 6 by a discussion of similar equations.) The aim, therefore, is to solve

$$u_t - 6uu_x + u_{xxx} = 0, \qquad -\infty < x < \infty, \quad t > 0,$$

for $u(x, t)$, with $u(x, 0)$ given. We anticipate that $u(x, 0)$ will have to satisfy some conditions in order that $u(x, t)$ exists, but hope that the conditions will be fairly weak. It will turn out (Chap. 4) that the method of solution requires a connection to be made between $u(x, t)$ and a scattering problem, in fact the classical scattering problem of quantum mechanics. It is this idea which is at the heart of the so-called *inverse scattering transform*.

Once this connection has been made it is then a matter of recalling the relevant information from the linear scattering and inverse scattering theories. To this end we digress in this chapter in order to present the results that we shall require later. We shall, however, discuss primarily the features relevant to soliton theory. Thus we shall emphasise the mathematical aspects, and only mention the quantum mechanical interpretation in passing. The development will revolve around the *Sturm–Liouville problem* on the whole line for the bounded function $\psi(x; \lambda)$, where

$$\psi_{xx} + (\lambda - u)\psi = 0, \qquad -\infty < x < \infty, \tag{3.1}$$

given a real *potential* function $u(x)$. Of course, this is just the time-independent Schrödinger equation in one dimension, and the *eigenvalue problem* defined by equation (3.1), for the parameter λ, is then the *scattering problem* of quantum mechanics. On the other hand, the *scattering data*, i.e. the form of $\psi(x; \lambda)$ as $x \to \pm\infty$, can determine uniquely the potential $u(x)$ which gave rise to these data; this is the *inverse scattering problem*. (There is good reason to use the symbol u for the potential in equation (3.1), as well as for the solution of the KdV equation, as we shall see later. At this stage we observe that they are different: one is $u(x)$ and the other is $u(x, t)$.) We shall now examine the direct and inverse scattering problems in detail.

3.2 The scattering problem

In order that appropriate solutions exist to the equation

$$\psi_{xx} + (\lambda - u)\psi = 0,$$

we shall require that $u(x)$ be integrable, i.e.

$$\int_{-\infty}^{\infty} |u(x)| \, dx < \infty.$$

In fact, $u(x)$ must decay sufficiently rapidly at infinity so that the Faddeev condition (Faddeev, 1958),

$$\int_{-\infty}^{\infty} (1 + |x|) |u(x)| \, dx < \infty, \tag{3.2}$$

is also satisfied. We shall not explain how these conditions arise because they involve subtle points of analysis that occur in the various existence theorems, and so are unimportant in the practical solution of problems. We shall, however, assume that they do hold and since we shall exhibit certain solutions explicitly the problem of existence is (more-or-less) avoided (and, as it happens, condition (3.2) is more stringent than we usually require). We shall further assume – for most problems – that $u(x)$ is infinitely differentiable, which is also not a necessary condition. (Potentials which are square wells or delta functions, for example, are commonly met at an undergraduate level.)

The spectrum of eigenvalues, λ, is generally made up of two types corresponding to $\lambda > 0$ and $\lambda < 0$ (and note that $\lambda = 0$ does not occur if $u(x) \neq 0$ for some x). Similarly there are two types of eigenfunction which can be easily characterised: since $u \to 0$ as $x \to \pm\infty$ then

$$\psi_{xx} \sim -\lambda\psi,$$

and so for all $\lambda > 0$ (the *continuous spectrum*) the eigenfunction ψ is (asymptotically) a linear combination of $\exp(\pm i\lambda^{1/2}x)$. If $\lambda < 0$, however, ψ will involve exponential growth or decay terms (or both). Let us consider the (real) solution which decays as $x \to -\infty$,

$$\psi(x) \sim \alpha e^{(-\lambda)^{1/2}x},$$

for some constant α; in general this solution will incorporate *both* the exponential terms at $+\infty$,

$$\psi(x) \sim \beta e^{(-\lambda)^{1/2}x} + \gamma e^{-(-\lambda)^{1/2}x} \qquad \text{as } x \to +\infty,$$

where β and γ are also constants. However, this solution is not bounded as $x \to +\infty$ unless $\beta = 0$: the values of λ for which $\beta = 0$ constitute the *discrete spectrum* of eigenvalues. Thus these eigenfunctions decay exponentially as $x \to \pm\infty$; those corresponding to $\lambda > 0$ obviously oscillate at infinity. For a given $u(x)$ there may be no discrete spectrum at all (as happens if $u \geqslant 0$, $-\infty < x < \infty$). Even if there is a discrete spectrum, the continuous eigenfunctions[†] may take a particularly simple form as can happen for special $u(x) < 0$, $-\infty < x < \infty$. It should be mentioned also that, if $u(x) \leqslant 0$ for all x and $u \to 0$ sufficiently rapidly at infinity, there is a *finite* number of discrete eigenvalues. Rather than given formal proofs of the above statements we shall show how they arise, via examples, later. We shall not be concerned with degenerate eigenvalues i.e. those for which there is more than one eigenfunction.

Let us now say something about the nature of the eigenfunctions for general x. From the original Sturm–Liouville equation, (3.1), we can see by direct integration that

$$[\psi_x]_b^a = \int_b^a (u - \lambda)\psi \, dx. \tag{3.3}$$

Thus, on taking the limit as $b \to a$ we see that ψ_x is *continuous* at every $x = a$ if the integral in equation (3.3) approaches zero in this limit. This is usually the case (and, of course, may be so even for discontinuous u), but if, for example, u is a delta function centred between a and b then more care is required in the calculation (see example (i) later). Similarly, we can first integrate equation (3.1) from c to x, and then from b to a to give

$$[\psi]_b^a = (a - b)\psi_x(c) + \int_b^a \left\{ \int_c^x (u - \lambda)\psi \, dx \right\} dx,$$

and so ψ is continuous everywhere for all functions u of interest, which includes the delta function. In other words, we shall be discussing eigenfunctions ψ which are bounded and continuous, and usually at least once differentiable.

Functions which are bounded and decay exponentially at infinity are integrable in a number of senses. Thus for the discrete eigenfunctions we can say that

$$\int_{-\infty}^{\infty} |\psi| \, dx < \infty \qquad \text{and} \qquad \int_{-\infty}^{\infty} |\psi|^2 \, dx = \int_{-\infty}^{\infty} \psi^2 \, dx < \infty,$$

[†] We shall use this short-hand to denote the eigenfunctions associated with the continuous spectrum, and similarly for the discrete spectrum.

where this second result means that $\psi(x)$ (which here is real) is 'square integrable'. The eigenfunctions associated with the continuous spectrum, however, are *not* square integrable (the corresponding integral is infinite) even though the ψs are bounded.

We are now in a position to introduce the most convenient representation and notation for the solutions relevant to our purpose. First, for the discrete spectrum, we define $\kappa_n = (-\lambda)^{1/2}$ for each of the discrete eigenvalues $(n = 1, 2, \ldots, N)$ and we shall order the eigenvalues according to the convention $\kappa_1 < \kappa_2 < \cdots < \kappa_N$. The bounded solution will then be characterised by its behaviour at $+\infty$ by writing

$$\psi_n(x) \sim c_n \exp(-\kappa_n x) \qquad \text{as } x \to +\infty, \tag{3.4}$$

where the subscript n denotes the nth eigenfunction. The real constant c_n is fixed by normalising the eigenfunction, via the square-integrability, so that

$$\int_{-\infty}^{\infty} \psi_n^2 \, dx = 1. \tag{3.5}$$

(Another choice would be to define the solution so that $\psi_n \sim \exp(-\kappa_n x)$ as $x \to +\infty$ – the *Jost solutions* – but (3.5) is more usually adopted in this context.) For the continuous spectrum we write $\sqrt{\lambda} = k$ and define the solution which is a special (linear) combination of the two oscillatory behaviours at infinity,

$$\hat{\psi}(x; k) \sim \begin{cases} e^{-ikx} + b e^{ikx} & \text{as } x \to +\infty, \\ a e^{-ikx} & \text{as } x \to -\infty. \end{cases} \tag{3.6}$$

Just as with c_n above, the two complex constants $a(k)$, $b(k)$ can be determined uniquely from a given $u(x)$. (Again, alternatives would be $\psi \sim e^{\pm ikx}$ as $x \to \pm \infty$, the Jost solutions for the continuous spectrum, which are useful in discussing the analytic properties of these eigenfunctions: see §3.3.)

If we now consider two different discrete eigenfunctions (for the same u, of course) then

$$\psi_n'' - (\kappa_n^2 + u)\psi_n = 0 \qquad \text{and} \qquad \psi_m'' - (\kappa_m^2 + u)\psi_m = 0,$$

(on reverting to primes to denote derivatives) and so

$$(\kappa_n^2 - \kappa_m^2)\psi_n \psi_m = \psi_m \psi_n'' - \psi_n \psi_m'' = \frac{d}{dx} W(\psi_m, \psi_n), \tag{3.7}$$

where we define $W(\alpha, \beta) = \alpha\beta' - \beta\alpha'$ as the *Wronskian* of α and β. Integration of equation (3.7) now yields

$$[W(\psi_m, \psi_n)]_{-\infty}^{\infty} = (\kappa_n^2 - \kappa_m^2) \int_{-\infty}^{\infty} \psi_n \psi_m \, dx,$$

and since $\psi_m, \psi_n \to 0$ as $x \to \pm \infty$ we obtain

$$\int_{-\infty}^{\infty} \psi_n \psi_m \, dx = 0 \qquad (m \neq n); \qquad (3.8)$$

i.e. the functions ψ_m, ψ_n are *orthogonal*. Furthermore, the continuous eigenfunction $\hat{\psi}$ is also orthogonal to every discrete eigenfunction, ψ_n (see Q3.2). (The discrete and continuous eigenfunctions together form a *complete set*, so that any square-integrable function can be represented as a linear combination of all the ψ_ns, plus an integral of $\hat{\psi}$ over all k.)

Equation (3.7) shows that, if θ and ϕ are two solutions of equation (3.1) with the same value of $\lambda = k^2$ (>0), then

$$\frac{d}{dx} W(\theta, \phi) = 0 \qquad \text{and so} \qquad W(\theta, \phi) = \text{constant.}$$

If, moreover, ϕ is proportional to θ, then $W(\theta, \phi) = 0$ for all x. On the other hand, if we introduce $\hat{\psi}$ (the continuous eigenfunction) and $\hat{\psi}^*$ (its complex conjugate), then $W(\hat{\psi}, \hat{\psi}^*)$ may be evaluated at both $\pm \infty$ (on using conditions (3.6)) to yield

$$W(\hat{\psi}, \hat{\psi}^*) = 2ikaa^* = 2ik \,(1 - bb^*),$$

or

$$|a|^2 + |b|^2 = 1. \qquad (3.9)$$

There is not much more that we can expect to learn without prescribing the particular form of $u(x)$. Once this is done, however, we can use standard techniques to solve the resulting second-order ordinary differential equation for $\psi(x)$. Before we explore these ideas further – and clarify some of the points made earlier – by examining two examples, we shall briefly relate these details to the quantum scattering problem. The discrete eigenfunctions are, of course, the *bound states* (sometimes called 'stationary' states). The eigenfunction $\hat{\psi}$, associated with the continuous spectrum, is interpreted as an *incident* wave of unit amplitude from $+\infty$ (e^{-ikx}), together with a *reflected* wave (be^{ikx}) to $+\infty$ and a *transmitted* wave (ae^{-ikx}) to

Fig. 3.1 Sketch representing the scattering by a potential.

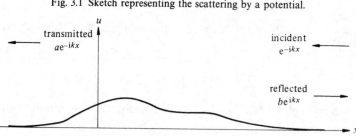

$-\infty$: see Fig. 3.1. Thus the constants a, b are usually referred to as the *transmission* and *reflection coefficients*, respectively. The Wronskian relation, (3.9), also has a meaning: it is a statement of the *conservation of energy* in the scattering. Many other details can be extracted, and they are of considerable interest in quantum mechanics, but are not relevant to our discussion.

We conclude this section with an analysis of two particular choices for $u(x)$: a delta function and a sech^2 function. The first of these gives a classical problem in quantum scattering theory, and the second plays an important rôle in our study of the KdV equation. We shall demonstrate how the eigenvalues and eigenfunctions can be determined for them both, and we shall also take the opportunity to explain what governs the number of discrete eigenvalues.

Example (i): the delta function

In this case we choose

$$u(x) = -U_0 \delta(x) \tag{3.10}$$

where U_0 is a constant, and $\delta(x)$ is Dirac's delta function. Before we can present the results we must first determine the nature of the solution at $x = 0$. The first integral, (3.3), of the Sturm–Liouville equation, from $-\varepsilon$ to $+\varepsilon$, is

$$[\psi']_{-\varepsilon}^{\varepsilon} + \int_{-\varepsilon}^{\varepsilon} \{\lambda + U_0 \delta(x)\} \psi(x)\, dx = 0$$

and so as $\varepsilon \to 0$,

$$[\psi'] = -U_0 \psi(0), \tag{3.11}$$

where $[\ ]$ denotes the change in ψ', provided ψ is continuous at $x = 0$. A second integration confirms that $\psi(x)$ is indeed continuous everywhere but, as we have just seen, it is not differentiable at $x = 0$.

For the discrete spectrum we set $\kappa_n = (-\lambda)^{1/2}$, and write

$$\psi_n(x) = \begin{cases} \alpha_n \exp(-\kappa_n x), & x > 0 \\ \beta_n \exp(\kappa_n x), & x < 0, \end{cases} \tag{3.12}$$

since $u = 0$ if $x \neq 0$; ψ_n is then continuous if $\alpha_n = \beta_n$. This solution is normalised by

$$\int_{-\infty}^{0} \alpha_n^2 \exp(2\kappa_n x)\, dx + \int_{0}^{\infty} \alpha_n^2 \exp(-2\kappa_n x)\, dx = 1$$

and so $\alpha_n = (\kappa_n)^{1/2}$ if we choose $\alpha_n > 0$. Finally, the discontinuity in ψ'_n, given by equation (3.11), requires that

$$[\psi_n'] = -\kappa_n \alpha_n - \kappa_n \alpha_n = -U_0 \alpha_n$$

and so $\kappa_n = \frac{1}{2}U_0$. Hence there is only one eigenvalue $(\lambda_1 = -\frac{1}{4}U_0^2;$ $\kappa_1 = \frac{1}{2}U_0)$, and then only if $U_0 > 0$; if $U_0 < 0$ there is none since $\alpha_n = (\kappa_n)^{1/2} = (U_0/2)^{1/2}$ must be real.

The continuous eigenfunction (for $k = \lambda^{1/2}$) can be written down from conditions (3.6) directly as

$$\hat{\psi}(x; k) = \begin{cases} e^{-ikx} + b(k)e^{ikx}, & x > 0 \\ a(k)e^{-ikx}, & x < 0 \end{cases} \tag{3.13}$$

which is continuous at $x = 0$ provided $a = 1 + b$. The discontinuity in $\hat{\psi}'$ at $x = 0$ is accommodated if

$$[\hat{\psi}'] = -ik + bik - (-ika) = -U_0(1 + b)$$

or

$$b(k) = -U_0/(U_0 + 2ik). \tag{3.14}$$

In this example the continuous spectrum always exists, but the discrete spectrum exists only if $U_0 > 0$ (and then it has only one member). Note further that the *pole* (in the upper half-plane) of $b(k)$ (and also of $a(k)$) corresponds to the discrete eigenvalue if k is extended into the complex plane i.e. at $k = iU_0/2$, $\lambda = -\frac{1}{4}U_0^2$. This is a quite general result (see §3.3) and enables all the discrete eigenvalues to be determined from the eigenfunction for the continuous spectrum.

It might be thought that a more natural example to introduce these ideas would be the rectangular-well potential

$$u(x) = \begin{cases} U_0, & 0 < x < 1 \\ 0, & x < 0, x > 1, \end{cases} \tag{3.15}$$

where U_0 is a constant. This choice does have the advantage that both ψ and ψ' are continuous everywhere (and in particular at $x = 0, 1$). However it turns out that the details are considerably more involved than for the delta function. The problem arising from the potential (3.15) is left as an exercise: see Q3.4.

Example (ii): the sech^2 function

The choice

$$u(x) = -U_0 \operatorname{sech}^2 x, \tag{3.16}$$

where U_0 is a constant, will be useful in our discussion of the KdV equation; it is also a problem we can solve explicitly for ψ (by using the hypergeometric function). The Sturm–Liouville equation is now

$$\psi'' + (\lambda + U_0 \operatorname{sech}^2 x)\psi = 0, \tag{3.17}$$

which is conveniently transformed by the substitution $T = \tanh x$ (so that $-1 < T < 1$ for $-\infty < x < \infty$). Thus

$$\frac{d}{dx} \equiv \operatorname{sech}^2 x \frac{d}{dT} = (1 - T^2)\frac{d}{dT}$$

and so

$$(1 - T^2)\frac{d}{dT}\left\{(1 - T^2)\frac{d\psi}{dT}\right\} + \{\lambda + U_0(1 - T^2)\}\psi = 0$$

or

$$\frac{d}{dT}\left\{(1 - T^2)\frac{d\psi}{dT}\right\} + \left(U_0 + \frac{\lambda}{(1 - T^2)}\right)\psi = 0, \tag{3.18}$$

which is the *associated Legendre equation*.

Let us first suppose that $U_0 = N(N + 1)$, where N is a positive integer, and consider the discrete spectrum. (More general U_0 will be mentioned later.) If $\lambda = -\kappa^2$ (< 0), then the only bounded solutions for $T \in [-1, 1]$ occur when $\kappa_n = n$, $n = 1, 2, \ldots, N$, and these solutions are proportional to the *associated Legendre functions*, $P_N^n(T)$, where

$$P_N^n(T) = (-1)^n(1 - T^2)^{n/2}\frac{d^n}{dT^n}P_N(T) \quad \text{and} \quad P_N(T) = \frac{1}{N!2^N}\frac{d^N}{dT^N}(T^2 - 1)^N, \tag{3.19}$$

$P_N(T)$ being the Legendre polynomial of degree N. (The proportionality constant is chosen to satisfy the normalisation condition for the discrete eigenfunctions, (3.5).) So, for example, if $N = 2$ we have the eigenfunctions

$$\psi_1(x) \propto P_2^1(\tanh x) = -3 \tanh x \operatorname{sech} x$$

and

$$\psi_2(x) \propto P_2^2(\tanh x) = 3 \operatorname{sech}^2 x,$$

since $P_2(T) = \frac{1}{2}(3T^2 - 1)$. If ψ_1 and ψ_2 are normalised, we obtain

$$\psi_1(x) = \left(\frac{3}{2}\right)^{1/2}\tanh x \operatorname{sech} x; \qquad \psi_2(x) = \frac{3^{1/2}}{2}\operatorname{sech}^2 x. \tag{3.20}$$

The corresponding eigenvalues are then $\lambda_1 = -1$, $\lambda_2 = -4$ ($\kappa_1 = 1$, $\kappa_2 = 2$).

Let us next consider the continuous spectrum, $\lambda = k^2$ (> 0): the solution of equation (3.18) which behaves like conditions (3.6) as $x \to -\infty$ can be written as

$$\hat{\psi}(x; k) = a(k)2^{ik}(\operatorname{sech} x)^{-ik}F(\tilde{a}, \tilde{b}; \tilde{c}; \tfrac{1}{2}(1 + T)), \tag{3.21}$$

where $\tilde{a} = \frac{1}{2} - ik + (U_0 + \frac{1}{4})^{1/2}$, $\tilde{b} = \frac{1}{2} - ik - (U_0 + \frac{1}{4})^{1/2}$ and $\tilde{c} = 1 - ik$. $F(\tilde{a}, \tilde{b}; \tilde{c}; z)$ is the hypergeometric function, for which $z \to 0^+$ corresponds to $x \to -\infty$, and $z \to 1^-$ to $x \to +\infty$, where $z = \frac{1}{2}(1 + T)$. It is a fairly simple exercise to confirm that

$$\hat{\psi}(x; k) \sim a(k) e^{-ikx} \qquad \text{as } x \to -\infty,$$

and we obtain

$$\hat{\psi}(x; k) \sim \frac{a\Gamma(\tilde{c})\Gamma(\tilde{a} + \tilde{b} - \tilde{c})}{\Gamma(\tilde{a})\Gamma(\tilde{b})} e^{-ikx} + \frac{a\Gamma(\tilde{c})\Gamma(\tilde{c} - \tilde{a} - \tilde{b})}{\Gamma(\tilde{c} - \tilde{a})\Gamma(\tilde{c} - \tilde{b})} e^{ikx} \qquad \text{as } x \to +\infty.$$

Comparing this with conditions (3.6) we therefore require that

$$a(k) = \frac{\Gamma(\tilde{a})\Gamma(\tilde{b})}{\Gamma(\tilde{c})\Gamma(\tilde{a} + \tilde{b} - \tilde{c})} \qquad \text{and} \qquad b(k) = \frac{a(k)\Gamma(\tilde{c})\Gamma(\tilde{c} - \tilde{a} - \tilde{b})}{\Gamma(\tilde{c} - \tilde{a})\Gamma(\tilde{c} - \tilde{b})}, \qquad (3.22)$$

which completely determines the scattering coefficients (see Q3.10).

From equations (3.22) we can derive two important and interesting results. The first of these, that there are certain U_0 for which $b(k) = 0$ for all k, is perhaps surprising. In terms of the scattering process this implies that all the incident wave is transmitted, and so these potentials are termed *reflectionless*. To see how this arises we need a standard identity relating gamma functions,

$$\Gamma(\tfrac{1}{2} - z)\Gamma(\tfrac{1}{2} + z) = \pi/\cos \pi z,$$

and hence observe that

$$\Gamma(\tilde{c} - \tilde{a})\Gamma(\tilde{c} - \tilde{b}) = \Gamma\{\tfrac{1}{2} - (U_0 + \tfrac{1}{4})^{1/2}\}\Gamma\{\tfrac{1}{2} + (U_0 + \tfrac{1}{4})^{1/2}\} = \pi/\cos\{\pi(U_0 + \tfrac{1}{4})^{1/2}\}.$$

Consequently $b(k) = 0$ for all k if

$$(U_0 + \tfrac{1}{4})^{1/2} = N + \tfrac{1}{2} \qquad \text{i.e. } U_0 = N(N + 1),$$

where N is a positive integer; thus the discrete eigenfunctions described by the associated Legendre functions belong to the class of reflectionless potentials.

The second result is concerned with the number of discrete eigenvalues for general U_0. Although we may examine the original equation, (3.18), directly, it is far simpler to make use of the poles (in the upper half-plane) of $a(k)$ and $b(k)$. These two coefficients have poles where $\Gamma(\tilde{a})$ or $\Gamma(\tilde{b})$ are undefined, and in the upper half-plane these occur where $\tilde{b} = -m$ $(m = 0, 1 \ldots)$, or

$$k = i\{(U_0 + \tfrac{1}{4})^{1/2} - (m + \tfrac{1}{2})\}.$$

Thus there is a finite number of discrete eigenvalues (for finite U_0) if

$$(U_0 + \tfrac{1}{4})^{1/2} > \tfrac{1}{2} \qquad \text{i.e. } U_0 > 0;$$

if $U_0 < 0$ there is no discrete eigenvalue. (The special case $U_0 = N(N+1)$ is easily confirmed, for now $k = i(N-m) = in$, $n = 1, 2, ..., N$.) The total number of eigenvalues, for general U_0, is obviously given by $[(U_0 + \frac{1}{4})^{1/2} - \frac{1}{2}] + 1$, where [ ] here denotes the integral part. Note, however, that if $\{(U_0 + \frac{1}{4})^{1/2} - \frac{1}{2}\}$ is integral then the $+1$ is omitted. (The eigenfunctions corresponding to

$$\kappa_m = (U_0 + \tfrac{1}{4})^{1/2} - (m + \tfrac{1}{2}) = \mu$$

are the associated Legendre functions $P_\nu^\mu(T)$, where ν is a root of $\nu(\nu+1) = U_0$.)

In conclusion, this example has introduced the reflectionless potentials and reinforced the general conditions which must pertain for the existence of discrete eigenvalues. In particular, if $U_0 > 0$, a finite number of such eigenvalues will arise and their values can be determined; if $U_0 < 0$, only the continuous spectrum is present. (When $U_0 < 0$, the corresponding eigenfunctions are given by solution (3.21) since this solution is valid for all real U_0.) It is also evident from this example that some knowledge and skill in the solution of second-order ordinary differential equations are assets in solving scattering problems!

*3.3 The inverse scattering problem

Eigenvalue problems of the type discussed in the previous section were fairly well understood by about 1850, but it was not until 1951 that the inverse problem was solved. In physical terms the problem is essentially one of finding the shape (or perhaps mass distribution) of an object which is mechanically vibrated, from a knowledge of all the sounds that it makes, i.e. from the energy or amplitude at each frequency. In our terms, for example, given $b(k)$ can we find $u(x)$? This is not so straightforward as the scattering problem (and it might even seem an impossible task!), but we shall give a reasonably complete picture of the theory without going into a lot of detail. All the information relevant to the ultimate aim of developing a solution of the KdV equation will, however, be presented. As with the scattering theory, the interested reader may pursue those points which are beyond the scope of this text. This and the next section are rather technical and so, at a first reading, you are recommended to skip directly to Chap. 4.

In order to motivate the initial phase of the calculation let us first consider the classical wave equation for $\phi(x, z)$,

$$\phi_{xx} - \phi_{zz} = 0. \tag{3.23}$$

As suggested in §1.1, this can be solved in many different ways; the method we adopt here is the use of the Fourier transform. If we write

$$\phi(x,z) = \frac{1}{2\pi} \int_{-\infty}^{\infty} \psi(x;k) e^{-ikz} \, dk \tag{3.24}$$

then

$$\psi(x;k) = \int_{-\infty}^{\infty} \phi(x,z) e^{ikz} \, dz \tag{3.25}$$

is the Fourier transform of $\phi(x,z)$. Hence we can easily see, upon the substitution of (3.24) into equation (3.23), that $\psi(x;k)$ must satisfy

$$\psi_{xx} + k^2 \psi = 0. \tag{3.26}$$

Thus the eigenfunction equation, (3.26), is the Fourier transform of the wave equation, (3.23). Further, let us suppose that we are interested in finding a solution of equation (3.26) for which $\psi \sim e^{ikx}$ as $x \to +\infty$. This can be accomplished if we choose

$$\phi(x,z) = \delta(x - z) + K(x,z),$$

where $K(x,z) = 0$ if $z < x$, and $K(x,z)$ is a solution of equation (3.23), for then

$$\psi(x;k) = e^{ikx} + \int_{x}^{\infty} K(x,z) e^{ikz} \, dz. \tag{3.27}$$

Now, our aim is to solve the inverse scattering problem for the equation

$$\psi_{xx} + (k^2 - u)\psi = 0,$$

with $\lambda = k^2$ (i.e. for the continuous spectrum) and where $u(x)$ is given. This Sturm–Liouville equation is similar to equation (3.26) and is, very importantly, linear in ψ. Moreover, we are seeking solutions for which $\psi \sim e^{ikx}$ as $x \to +\infty$ (which describes the Jost solutions for real k). We consider, first, potentials for which there is no discrete spectrum. To proceed, therefore, we may seek a solution of the Sturm–Liouville equation which takes the form (3.27) for some $K(x,z)$. We can anticipate that the equation for $K(x,z)$ is a wave equation with an additional term (perhaps uK?), although the connection with u is not clear at present.

If the solution is assumed to be

$$\psi_{+}(x;k) = e^{ikx} + \int_{x}^{\infty} K(x,z) e^{ikz} \, dz, \tag{3.28}$$

where the subscript $+$ denotes a solution satisfying conditions as $x \to +\infty$, then a direct calculation shows that

$$\psi_{+xx} = e^{ikx}\left(-k^2 - \frac{d\hat{K}}{dx} - ik\hat{K} - \hat{K}_x \right) + \int_{x}^{\infty} K_{xx} e^{ikz} \, dz$$

where $\hat{K}(x) = K(x,x)$, the subscripts denote partial derivatives and d/dx

is a total derivative i.e. $d\hat{K}/dx = \hat{K}_x + \hat{K}_z$, where the circumflex over the partial derivative indicates evaluation on $z = x$. Also, ψ_+ itself can be conveniently written as

$$\psi_+ = e^{ikx}\left(1 + \frac{i\hat{K}}{k} - \frac{\hat{K}_z}{k^2}\right) - \frac{1}{k^2}\int_x^\infty K_{zz}e^{ikz}\,dz,$$

after integration by parts twice, provided both $K(x, z)$ and $K_z(x, z) \to 0$ as $z \to +\infty$ for fixed x. The Sturm–Liouville equation now becomes

$$0 = \psi_{+xx} + (k^2 - u)\psi_+$$
$$= -e^{ikx}\left(u + 2\frac{d\hat{K}}{dx}\right) + \int_x^\infty (K_{xx} - K_{zz} - u(x)K)e^{ikz}\,dz$$

which is satisfied if

$$K_{xx} - K_{zz} - u(x)K = 0 \qquad (z > x) \tag{3.29}$$

and

$$u(x) = -2\frac{d\hat{K}}{dx} = -2\{K_x(x, x) + K_z(x, x)\}, \tag{3.30}$$

which together with the conditions

$$K(x, z), K_z(x, z) \to 0 \qquad \text{as } z \to +\infty \tag{3.31}$$

defines the problem for $K(x, z)$. Thus, given $u(x)$, $K(x, z)$ is a real-valued function which satisfies a wave equation (equation (3.29) where x or z plays the rôle of time), with data given on both $z = x$ (from equation (3.30)) and as $z \to +\infty$. Since $z = x$ is one of the characteristic lines of equation (3.29), this constitutes a *Goursat problem*. It is well-known that the solution exists and is unique (see e.g. Garabedian, 1964), and indeed $K(x, z)$ can be completely determined using Riemann's method (but this will profit us little here).

At this stage it is apparent that we are still developing the direct scattering problem since we are viewing $K(x, z)$ as a function to be found if $u(x)$ is specified. The crucial point now is that we can investigate the possibility of inverting equation (3.28) (or something like it) to obtain $K(x, z)$ from a known $\psi_+(x; k)$, and hence use equation (3.30) to find $u(x)$. This procedure is allowable because we have just shown that K exists. Before we pursue this line, a small matter of consistency is worth noting: it is clear from equations (3.29)–(3.31) that K is independent of k, and hence inversion of equation (3.28) is meaningful in the sense of transforms. Of course, this merely confirms the interpretation that $\psi_+(x; k)$ is the Fourier transform with respect to z of $\delta(x - z) + K(x, z)$.

The eigenfunction for the continuous spectrum is $\hat{\psi}$ (which is to satisfy conditions (3.6) at infinity as well as equation (3.1)) and we can construct $\hat{\psi}$ from the function ψ_+ by writing

$$\hat{\psi} = \psi_+^* + b(k)\psi_+ \tag{3.32}$$

which gives the correct behaviour as $x \to +\infty$. (We could equally well develop the above argument for $x \to -\infty$ by suitably altering equation (3.28), and using the behaviour as $x \to -\infty$ from conditions (3.6); both approaches give the same result.) From equation (3.32), and the solution (3.28) for ψ_+, we obtain

$$\hat{\psi}(x; k) = \mathrm{e}^{-\mathrm{i}kx} + b(k)\mathrm{e}^{\mathrm{i}kx} + \int_{-\infty}^{\infty} K(x, z)\mathrm{e}^{-\mathrm{i}kz}\,\mathrm{d}z + b(k)\int_{-\infty}^{\infty} K(x, z)\mathrm{e}^{\mathrm{i}kz}\,\mathrm{d}z$$

since $K(x, z) = 0$ for $z < x$. Upon re-arranging this equation to give

$$\int_{-\infty}^{\infty} K(x, z)\mathrm{e}^{-\mathrm{i}kz}\,\mathrm{d}z = \hat{\psi} - \mathrm{e}^{-\mathrm{i}kx} - b(k)\mathrm{e}^{\mathrm{i}kx} - b(k)\int_{-\infty}^{\infty} K(x, z)\mathrm{e}^{\mathrm{i}kz}\,\mathrm{d}z,$$

we can invert both sides to yield

$$K(x, z) = \frac{1}{2\pi}\int_{-\infty}^{\infty} \left\{ \hat{\psi}(x; k) - \mathrm{e}^{-\mathrm{i}kx} - b(k)\mathrm{e}^{\mathrm{i}kx} \right.$$

$$\left. - b(k)\int_{-\infty}^{\infty} K(x, y)\mathrm{e}^{\mathrm{i}ky}\,\mathrm{d}y \right\} \mathrm{e}^{\mathrm{i}kz}\,\mathrm{d}k, \tag{3.33}$$

where y replaces z as the variable for the inner integration. Note that we have here the symmetry relations:

$$\psi(x; k) = \int_{-\infty}^{\infty} K(x, z)\mathrm{e}^{\mathrm{i}kz}\,\mathrm{d}z; \qquad \psi(x; -k) = \int_{-\infty}^{\infty} K(x, z)\mathrm{e}^{-\mathrm{i}kz}\,\mathrm{d}z$$

and

$$K(x, z) = \frac{1}{2\pi}\int_{-\infty}^{\infty} \psi(x; k)\mathrm{e}^{-\mathrm{i}kz}\,\mathrm{d}k = \frac{1}{2\pi}\int_{-\infty}^{\infty} \psi(x; -k)\mathrm{e}^{\mathrm{i}kz}\,\mathrm{d}k.$$

It is now convenient to define a function F by

$$F(x) = \frac{1}{2\pi}\int_{-\infty}^{\infty} b(k)\mathrm{e}^{\mathrm{i}kx}\,\mathrm{d}k, \tag{3.34}$$

which is an integral along the real k-axis (where $b(k)$ is an entire function in the absence of any discrete eigenvalues). From equations (3.33) and (3.34), after interchanging the order of integration in the double integral, we obtain

$$K(x, z) = \frac{1}{2\pi}\int_{-\infty}^{\infty} (\hat{\psi} - \mathrm{e}^{-\mathrm{i}kx})\mathrm{e}^{\mathrm{i}kz}\,\mathrm{d}k - F(x + z) - \int_{-\infty}^{\infty} K(x, y)F(y + z)\,\mathrm{d}y.$$

$$\tag{3.35}$$

The first integral (on the right-hand side) can be evaluated by introducing the contour, C, in the upper-half of the k-plane comprised of a semicircular arc of radius R and the real line segment $[-R, R]$; see Fig. 3.2. The eigenfunction $\hat{\psi}$ has no poles in the upper half-plane and $|\hat{\psi}e^{ikx}| \to 1$ as $|k| \to \infty$ (since it can be shown that $|a| \to 1$ in this limit; see Q3.10). Furthermore, on writing the integrand as

$$(\hat{\psi}e^{ikx} - 1)e^{ik(z-x)}, \tag{3.36}$$

it is clear that this decays exponentially as $R \to \infty$ on the arc since $\mathscr{I}(k) > 0$ there, and $z > x$. From Cauchy's theorem, the integral of expression (3.36) around C is zero and the contribution along the semicircular arc clearly tends to zero as $R \to \infty$. Hence

$$\int_{-\infty}^{\infty} (\hat{\psi} - e^{-ikx})e^{ikz}\,\mathrm{d}z = 0$$

and so we obtain

$$K(x,z) + F(x+z) + \int_{x}^{\infty} K(x,y)F(y+z)\mathrm{d}y = 0, \qquad z > x > -\infty, \tag{3.37}$$

called the *Gel'fand–Levitan equation*, or more usually when written in this form the *Marchenko equation* for $K(x,z)$, given the function $F(x)$. Equation (3.37) is a *linear Fredholm integral equation* which can be solved in closed form for suitable F, and for which existence and uniqueness of K (already implied by equations (3.29)–(3.31)) can be confirmed by constructing the Neumann-expansion solution (see §3.4). We shall discuss the Marchenko equation, and its method of solution, more fully in the next section. The potential function, $u(x)$, is then obtained from equation (3.30).

To complete this analysis we must examine the unresolved problem of

Fig. 3.2 The contour C in the complex k-plane.

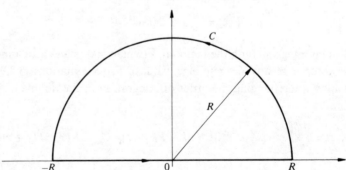

the discrete spectrum. It is fairly easy to confirm that if the discrete spectrum exists then the formulation of equation (3.35) is unaltered, with k now a complex parameter (although $F(x)$ given by equation (3.34) is still an integral along the real axis, of course). The difference arises in the evaluation of the integral containing $\hat{\psi}$, because the expression (3.36) will now have poles in the upper half-plane corresponding to the poles of $a(k)$. We evaluate this integral by using the contour given in Fig. 3.2 and ensuring that R is large enough to enclose all the poles. It then follows immediately from Cauchy's residue theorem that

$$\int_{-\infty}^{\infty} (\hat{\psi} - e^{-ikx})e^{ikz}\,dk = 2\pi i \sum_{n=1}^{N} R_n, \tag{3.38}$$

where R_n is the residue of $\hat{\psi}e^{ikz}$ at $k = i\kappa_n$ (on noting that $e^{ik(z-x)}$ has no poles). In order to obtain R_n we shall need to introduce the other Jost solution, and derive some identities involving Wronskians.

We define the solution

$$\psi_-(x;k) = e^{-ikx} + \int_{-\infty}^{x} L(x,z)e^{-ikz}\,dz, \tag{3.39}$$

for some appropriate $L(x,z)$ (whose existence is equivalent to that of $K(x,z)$: see Q3.11) so that $\psi_-(x;k) \sim e^{-ikx}$ as $x \to -\infty$. Thus we can write

$$\hat{\psi} = a(k)\psi_- = \psi_+^* + b(k)\psi_+ \tag{3.40}$$

and so

$$\psi_- = a^{-1}\psi_+^* + ba^{-1}\psi_+, \tag{3.41}$$

which is an important identity since both ψ_+ and ψ_- are *defined* at $k = i\kappa_n$. In fact, from definitions (3.28) and (3.39) we see that we have Jost solutions $\psi_+(x;i\kappa_n) \sim \exp(-\kappa_n x)$ as $x \to +\infty$; $\psi_-(x;i\kappa_n) \sim \exp(\kappa_n x)$ as $x \to -\infty$ for the discrete eigenfunctions. In particular we may therefore write

$$\psi_n(x) = c_n\psi_+(x;i\kappa_n) = d_n\psi_-(x;i\kappa_n), \tag{3.42}$$

where d_n is some (real) constant at present unknown, and c_n is the normalisation constant; see equations (3.4) and (3.5). (Note that, with $k = i\kappa_n$, the integrals in both equations (3.28) and (3.39) now become essentially Laplace transforms.)

The Wronskian $W(\psi_-, \psi_+)$ is a constant (see equation (3.7) et seq.) which can be evaluated by letting $x \to +\infty$ (and on using equation (3.41)), to yield

$$W(\psi_-, \psi_+) = 2ika^{-1}. \tag{3.43}$$

(This result may be used to prove that a^{-1} has zeros at $k = i\kappa_n$ for $n = 1, 2, \ldots, N$ (see Q3.12), and hence $a(k)$ has poles at these points.) The derivative of equation (3.43) with respect to k (if it is assumed that ψ_{+k}

and ψ_{-k} exist: see equations (3.28), (3.39)) is

$$W(\psi_{-k}, \psi_+) + W(\psi_-, \psi_{+k}) = 2ia^{-1} - 2ik(a'/a^2) \qquad (3.44)$$

where $a' \equiv da/dk$ (and equation (3.44) can now be invoked to show that $a(k)$ has only *simple* poles; see Q3.12). Furthermore, from the original Sturm–Liouville equation we can see, upon differentiating with respect to k, that

$$\psi_{xxk} + 2k\psi + (k^2 - u)\psi_k = 0$$

which, when multiplied by ψ, can be combined with

$$\psi_k \psi_{xx} + (k^2 - u)\psi\psi_k = 0$$

to yield

$$\psi_{xx}\psi_k - \psi\psi_{xxk} = 2k\psi^2$$

or

$$W(\psi_k, \psi) = 2k \int^x \psi^2 \, dx. \qquad (3.45)$$

In this last result, let us choose $\psi \equiv \psi_+$, set $k = i\kappa_n$ and take the integral from $-\infty$ to $+\infty$ so that

$$W_n^+(\psi_{+k}, \psi_+) - W_n^-(\psi_{+k}, \psi_+) = 2i\kappa_n \int_{-\infty}^{\infty} \psi_+^2 \, dx = \frac{2i\kappa_n}{c_n^2}, \qquad (3.46)$$

since $\psi_+(x; i\kappa_n) = \psi_n/c_n$ and ψ_n is normalised. The notation $W_n^\pm$ indicates evaluation at $\pm\infty$, and at $k = i\kappa_n$. Evaluating equation (3.44) as $x \to -\infty$, and with $k = i\kappa_n$, yields

$$W_n^-(\psi_{-k}, d_n\psi_-/c_n) + W_n^-(c_n\psi_+/d_n, \psi_{+k}) = 2\kappa_n(a'/a^2)_n, \qquad (3.47)$$

where equations (3.42) have been incorporated, and ( )$_n$ again denotes evaluation at $k = i\kappa_n$. Combining equations (3.46) and (3.47), we eliminate $W_n^-(\psi_+, \psi_{+k})$ (whose value is unknown) to give

$$W_n^+(\psi_{+k}, \psi_+) - \frac{d_n^2}{c_n^2} W_n^-(\psi_{-k}, \psi_-) = \frac{2i\kappa_n}{c_n^2} - \frac{2d_n\kappa_n}{c_n}(a'/a^2)_n, \qquad (3.48)$$

where we note that $W(\phi, \psi) = -W(\psi, \phi)$, and also $W(\alpha\phi, \beta\psi) = \alpha\beta W(\phi, \psi)$ if α, β are constants. The left-hand side of equation (3.48) may be evaluated directly from definitions (3.28) and (3.39) as zero, and so

$$(a'/a^2)_n = i(d_n c_n)^{-1}. \qquad (3.49)$$

The determination of the residues of $\hat{\psi} e^{ikz}$ (see equation (3.38)) is now reduced to a straightforward exercise if we employ the above results. From equations (3.41) and (3.42) we are able to write

$$\hat{\psi}e^{ikz} = a(k)\psi_-(x;k)e^{ikz} = \frac{c_n}{d_n}a\psi_+e^{ikz},$$

and only $a(k)$ has any poles. The required residue at $k = i\kappa_n$ is therefore

$$R_n = \frac{c_n}{d_n}\psi_+(x;i\kappa_n)\exp(-\kappa_n z) \times \{\text{residue of } a(k)\}$$

and since $a(k)$ has only simple poles we obtain

$$\text{residue of } a(k) \text{ (at } k = i\kappa_n) = -(a^2/a')_n$$

and so, from equation (3.49),

$$R_n = ic_n^2\psi_+(x;i\kappa_n)\exp(-\kappa_n z).$$

Note that the value of d_n is not required. From equation (3.28) we immediately have an expression for $\psi_+(x;i\kappa_n)$, and so

$$\int_{-\infty}^{\infty}(\hat{\psi} - e^{-ikx})e^{ikz}\,dz$$

$$= -2\pi\sum_{n=1}^{N} c_n^2\exp(-\kappa_n z)\left\{\exp(-\kappa_n x) + \int_{x}^{\infty} K(x,y)\exp(-\kappa_n y)\,dy\right\},$$

where the integration variable on the right is chosen as y to avoid any confusion. If this expression is now used in equation (3.35), we recover the same Marchenko equation

$$K(x,z) + F(x+z) + \int_{x}^{\infty} K(x,y)F(y+z)\,dy = 0 \qquad (3.50)$$

provided that we redefine F by

$$F(X) = \sum_{n=1}^{N} c_n^2\exp(-\kappa_n X) + \frac{1}{2\pi}\int_{-\infty}^{\infty} b(k)e^{ikX}\,dk, \qquad (3.51)$$

which in turn recovers expression (3.34) if there is no discrete spectrum (i.e. $c_n = 0$, $n = 1, \ldots, N$).

This completes the formulation of the inverse scattering problem, and in any future applications we shall just quote equations (3.50), (3.51). These demonstrate that, given the scattering data comprised of the reflection coefficient $b(k)$, the discrete eigenvalues, $-\kappa_n^2$, and the corresponding normalisation constants, c_n, then $u(x)$ can be completely determined from $K(x,z)$. This final stage, which requires the solution of the Marchenko equation, will now be examined briefly in conjunction with two illustrative examples.

*3.4 The solution of the Marchenko equation

The Marchenko equation,

$$K(x, z) + F(x + z) + \int_x^\infty K(x, y)F(y + z)\,dy = 0, \qquad (3.52)$$

as we have already mentioned, is a Fredholm integral equation. This is clear if we write the equation as

$$K(x, z) + F(x + z) + \int_{-\infty}^\infty K(x, y)F(y + z)\,dy = 0,$$

with $K(x, z) = 0$ for $z < x$; note that x plays the rôle of a parameter here. One direct – but formal – method of solving equation (3.52) for general $F(x)$ makes use of an iterative procedure. First, let us define

$$K_1(x, z) = \begin{cases} -F(x + z), & z > x \\ 0 & z < x \end{cases}$$

and

$$K_2(x, z) = -F(x + z) - \int_{-\infty}^\infty K_1(x, y)F(y + z)\,dy,$$

$$K_3(x, z) = -F(x + z) - \int_{-\infty}^\infty K_2(x, y)F(y + z)\,dy,$$

and so on. If $K_n(x, z) \to K(x, z)$ pointwise as $n \to \infty$ then the existence of the solution is confirmed. The infinite expansion so obtained is the *Neumann series*, and this constitutes a formal solution of the integral equation (3.52). However, it is unlikely that this construction will yield any simple closed-form solutions (which would obviously be of far more use to us). Of course, F might have a sufficiently simple form for a solution to be derived directly by the use of elementary techniques: see example (i) below.

It so happens that an important special case will arise in our discussion of the KdV equation which reduces equation (3.52) to a standard problem. Let us suppose that $F(x + z)$ is a *separable* function, so that

$$F(x + z) = \sum_{n=1}^N X_n(x)Z_n(z),$$

where N is finite. The Marchenko equation can therefore be written as

$$K(x, z) + \sum_{n=1}^N X_n(x)Z_n(z) + \sum_{n=1}^N Z_n(z) \int_x^\infty K(x, y)X_n(y)\,dy = 0$$

from which it is clear that the solution must take the form

$$K(x, z) = \sum_{n=1}^{N} L_n(x) Z_n(z).$$

Upon this substitution for $K(x, z)$ we obtain, for $n = 1, 2, \ldots, N$,

$$L_n(x) + X_n(x) + \sum_{m=1}^{N} L_m(x) \int_x^\infty Z_m(y) X_n(y) \, dy = 0, \qquad (3.53)$$

which are N *algebraic* equations for the N unknown functions, $L_n(x)$. The solution of this system is now straightforward, and at this stage it is left as an exercise for the reader. We shall, however, present a more detailed discussion in the next chapter where the integral in equation (3.53) is evaluated for the particular choice of $F(x)$ which arises in the discussion of the KdV equation. A simplified version of equation (3.53), with just two unknown functions, will nevertheless be included as an example in this section.

We now examine the two illustrative examples which will clarify the points mentioned above, and show in detail how the potential function is determined.

Example (i): reflection coefficient with one pole

Here we take the scattering data as

$$b(k) = -\beta/(\beta + ik); \qquad \psi(x) \sim \beta^{1/2} e^{-\beta x} \qquad \text{as } x \to +\infty,$$

where $\beta > 0$ is a constant. There is just one discrete eigenvalue, since $\beta > 0$, at $k = i\beta$ (so that $\kappa_1 = \beta$), with $c_1 = \beta^{1/2}$. From the definition (3.51), for F, we see that

$$F(X) = \beta e^{-\beta X} - \frac{\beta}{2\pi} \int_{-\infty}^\infty \frac{e^{ikX}}{\beta + ik} \, dk.$$

The value of the integral is easily found: first consider the contour C as given in Fig. 3.2 with $R > \beta$. Since $\mathscr{I}(k) > 0$ on the semicircular arc, the value of the line integral on this part of C approaches zero as $R \to \infty$ if $X > 0$. Thus by Cauchy's residue theorem we obtain

$$\int_{-\infty}^\infty \frac{e^{ikX}}{\beta + ik} \, dk = 2\pi e^{-\beta X}, \qquad X > 0.$$

If $X < 0$, consider a similar contour in the lower half-plane; it immediately follows, by Cauchy's integral theorem, that

$$\int_{-\infty}^\infty \frac{e^{ikX}}{\beta + ik} \, dk = 0, \qquad X < 0.$$

Thus $F(X)$ becomes simply

$$F(X) = \beta e^{-\beta X} H(-X),$$

where $H(x)$ is the Heaviside step function ($H(x) = 1$ if $x > 0$; $H(x) = 0$ if $x < 0$), and so from equation (3.52) we shall have

$$K(x, z) = 0 \qquad \text{for } x + z > 0.$$

However, if $x + z < 0$, equation (3.52) now reads

$$K(x, z) + \beta e^{-\beta(x+z)} + \beta \int_x^{-z} K(x, y) e^{-\beta(y+z)} \, dy = 0,$$

on remembering that $F(y + z) = 0$ if $y + z > 0$. One way[†] to obtain the solution of this equation is to integrate by parts to yield

$$K(x, z) + \beta e^{-\beta(x+z)} + K(x, x) e^{-\beta(x+z)} - K(x, -z)$$
$$+ \int_x^{-z} K_y(x, y) e^{-\beta(y+z)} \, dy = 0.$$

It is now clear that one solution of this equation is $K(x, z) = -\beta$, and this must be the required solution by virtue of uniqueness. Thus we have

$$K(x, z) = -\beta H(-x - z) \qquad \text{and so} \qquad K(x, x) = -\beta H(-2x).$$

The required potential function is therefore

$$u(x) = 2\beta \frac{d}{dx} H(-2x) = -2\beta \delta(x)$$

since $H(-2x)$ can be replaced by $H(-x)$, $H'(x) = \delta(x)$ and $\delta(x)$ is an even function. This recovers the potential function (3.10) used in example (i), §3.2, if we set $\beta = U_0/2$.

Example (ii): zero reflection coefficient

For this example, we shall suppose that we have two discrete eigenvalues such that

$$\psi_1(x) \sim c_1 \exp(-\kappa_1 x); \qquad \psi_2(x) \sim c_2 \exp(-\kappa_2 x) \qquad \text{as } x \to +\infty,$$

where $\kappa_1 \neq \kappa_2$ and $b(k) = 0$ for all k. From definition (3.51) we therefore obtain

$$F(X) = c_1^2 \exp(-\kappa_1 X) + c_2^2 \exp(-\kappa_2 X), \tag{3.54}$$

---

† Another way is to differentiate with respect to z: this is left as an exercise.

and so the Marchenko equation, (3.52), becomes

$$K(x,z) + c_1^2 \exp\{-\kappa_1(x+z)\} + c_2^2 \exp\{-\kappa_2(x+z)\}$$
$$+ \int_x^\infty K(x,y)[c_1^2 \exp\{-\kappa_1(y+z)\} + c_2^2 \exp\{-\kappa_2(y+z)\}]\,dy = 0.$$

$$(3.55)$$

The form of $F(x+z)$, from equation (3.54), shows that F is separable and therefore we set

$$K(x,z) = L_1(x)\exp(-\kappa_1 z) + L_2(x)\exp(-\kappa_2 z).$$

From equation (3.55) it follows that $L_1(x)$, $L_2(x)$ must satisfy

$$L_1 + c_1^2 \exp(-\kappa_1 x)$$
$$+ c_1^2\left[L_1 \int_x^\infty \exp(-2\kappa_1 y)\,dy + L_2 \int_x^\infty \exp\{-(\kappa_1+\kappa_2)y\}\,dy\right] = 0,$$
$$L_2 + c_2^2 \exp(-\kappa_2 x)$$
$$+ c_2^2\left[L_1 \int_x^\infty \exp\{-(\kappa_1+\kappa_2)y\}\,dy + L_2 \int_x^\infty \exp(-2\kappa_2 y)\,dy\right] = 0,$$

or, upon evaluating the integrals,

$$L_n + c_n^2 \exp(-\kappa_n x) + c_n^2 \sum_{m=1}^{2} \frac{L_m \exp\{-(\kappa_1+\kappa_2)x\}}{\kappa_m+\kappa_n} = 0, \qquad (3.56)$$

for $n = 1, 2$.

This system is conveniently written as

$$AL + B = 0,$$

where L and B are column vectors with elements L_n and $B_n = c_n^2 \exp(-\kappa_n x)$, respectively, and A is the square matrix with elements

$$A_{mn} = \delta_{mn} + c_m^2 \frac{\exp\{-(\kappa_m+\kappa_n)x\}}{\kappa_m+\kappa_n},$$

where δ_{mn} is the Kronecker delta. (This formalism, therefore, immediately allows the extension to N discrete eigenvalues: see Chap. 4.) The solution for L is

$$L = -A^{-1}B$$

and then $K(x,x) = E^T L$, where E is the column vector with elements $E_n = \exp(-\kappa_n x)$. The important simplifying observation is to note that

$$\frac{d}{dx} A_{mn} = -c_m^2 \exp\{-(\kappa_m+\kappa_n)x\} = -B_m E_n$$

and so

$$K = E_m L_m = -E_m A_{mn}^{-1} B_n = A_{mn}^{-1} \frac{\mathrm{d}}{\mathrm{d}x} A_{nm}, \tag{3.57}$$

when written in subscripted variables and using the summation convention. However, expression (3.57) is more readily recognised in the form

$$K(x,x) = \mathrm{tr}\left(A^{-1}\frac{\mathrm{d}A}{\mathrm{d}x}\right) = \frac{1}{|A|}\frac{\mathrm{d}|A|}{\mathrm{d}x} = \frac{\mathrm{d}}{\mathrm{d}x}\log|A|,$$

where $|A|$ denotes $\det A$, and so

$$u(x) = -2\frac{\mathrm{d}^2}{\mathrm{d}x^2}\log|A|. \tag{3.58}$$

For the two discrete eigenvalues we started with, then

$$|A| = \left\{1 + \frac{c_1^2}{2\kappa_1}\exp(-2\kappa_1 x)\right\}\left\{1 + \frac{c_2^2}{2\kappa_2}\exp(-2\kappa_2 x)\right\}$$
$$- \frac{c_1^2 c_2^2}{(\kappa_1 + \kappa_2)^2}\exp\left\{-2(\kappa_1 + \kappa_2)x\right\}. \tag{3.59}$$

Thus equation (3.58), with (3.59), is the required potential function. (To check the form of this result, let us retain only one eigenvalue by setting $c_2 = 0$. From equations (3.59) and (3.58) it is now straightforward to show that

$$u(x) = -\frac{4\kappa_1 c_1^2 \exp(-2\kappa_1 x)}{\left\{1 + \frac{c_1^2}{2\kappa_1}\exp(-2\kappa_1 x)\right\}^2}$$
$$= -2\kappa_1^2 \operatorname{sech}^2(\kappa_1 x + x_0),$$

where $\exp(x_0) = (2\kappa_1)^{1/2}/c_1$, which is one of the class of reflectionless potentials; see example (ii), §3.2.)

Further reading

3.2 The mathematical aspects related to second-order ordinary differential equations, particularly Sturm–Liouville theory, can be found in Ince (1927). For further details concerning quantum scattering theory see Landau & Lifshitz (1977), or any other standard physics text.

3.3. The inverse scattering problem is discussed in Gel' fand & Levitan (1951), Marchenko (1955), Kay & Moses (1956). Information concerning the hypergeometric function, and also the gamma function, is given in Abramowitz & Stegun (1964).

3.4 Further study of integral equations – a large subject – can be undertaken from any standard text e.g. Courant & Hilbert (1953).

Other approaches to the formulation of the inverse scattering problem, with particular reference to the KdV equation, may be found in Lamb (1980), Ablowitz & Segur (1981), Calogero & Degasperis (1982), Dodd, Eilbeck, Gibbon & Morris (1982).

Exercises

Q3.1 Let y_1, y_2 be two solutions of the differential equation

$$y'' + p(x)y' + q(x)y = 0.$$

Define the Wronskian W, of y_1 and y_2, and show that

$$W' + p(x)W = 0.$$

Hence deduce that either $W = 0$ for all x or W never vanishes (provided x and $p(x)$ remain finite).

Q3.2 Show that a continuous eigenfunction, $\hat{\psi}$, of equation (3.1), and any discrete eigenfunction, ψ_n, are orthogonal.

Q3.3 Find, if they exist, the eigenvalues and eigenfunctions of

$$\psi'' + (\lambda - U_0)\psi = 0,$$

where U_0 is any real constant.

Q3.4 *Two classical scattering problems.* Find the eigenvalues and eigenfunctions of

$$\psi'' + \{\lambda - u(x)\}\psi = 0$$

in these two cases:

(i) $$u(x) = \begin{cases} U_0, & 0 < x < 1, \\ 0, & x < 0, x > 1, \end{cases}$$

where U_0 is any real constant.

*(ii) $u(x) = -U_0\delta(x) - U_1\delta(x-1)$,

where U_0 and U_1 are positive constants, and show that there is only one discrete eigenfunction if $(U_0 + U_1)/(U_0 U_1) > 1$.

Q3.5 *Another scattering problem.* Find the eigenvalues and eigenfunctions of

$$\psi'' + \{\lambda - u(x)\}\psi = 0,$$

if $u(x)$ is the step potential

$$u(x) = \begin{cases} 0, & x < 0 \\ U_0, & x > 0, \end{cases}$$

where U_0 is a positive constant. Show, in particular, that there is a continuous eigenfunction, no discrete eigenfunction, and an eigenfunction which decays as $x \to +\infty$ but is oscillatory in $x < 0$.

Q3.6 Relate the scattering problem with the potential $u(x) = -U_0 \operatorname{sech}^2 \beta x$, for some positive constant β, to the problem discussed in example (ii), §3.2.

Q3.7 *A reflectionless potential.* Find the discrete eigenfunctions for $N = 3$, where $u(x) = -N(N+1) \operatorname{sech}^2 x$ (see example (ii), §3.2).

Q3.8 Small-perturbation scattering theory. If

$$\psi'' + \{\lambda - u(x)\}\psi = 0,$$

with $u(x) = -\varepsilon f(x)$, such that $0 \leqslant \int_{-\infty}^{\infty} f(x)\,dx < \infty$, discuss both the continuous and discrete spectra as follows.

(i) *Continuous spectrum:* With $\lambda = k^2$ and

$$\hat{\psi}(x;k) \sim \begin{cases} ae^{-ikx}, & x \to -\infty \\ e^{-ikx} + be^{ikx}, & x \to +\infty \end{cases}$$

deduce that

$$a(k;\varepsilon) \sim 1 + \frac{i\varepsilon}{2k} \int_{-\infty}^{\infty} f(x)\,dx$$

and

$$b(k;\varepsilon) \sim \frac{i\varepsilon}{2k} \int_{-\infty}^{\infty} e^{-2ikx} f(x)\,dx$$

as $\varepsilon \to 0$ for fixed k.

(ii) *Discrete spectrum:* Show that, if $\int_{-\infty}^{\infty} f(x)\,dx > 0$ and

$$\psi(x) \sim \begin{cases} \gamma e^{\kappa x}, & x \to -\infty \\ e^{-\kappa x}, & x \to +\infty, \end{cases}$$

then there is just one eigenfunction, and hence show that the corresponding eigenvalue is

$$\kappa = \kappa_1 \sim \tfrac{1}{2}\varepsilon \int_{-\infty}^{\infty} f(x)\,dx \qquad \text{as} \qquad \varepsilon \to 0^+.$$

Show, however, that if $\varepsilon \to 0^-$ then no discrete eigenvalue exists. If $\int_{-\infty}^{\infty} f(x)\,dx = 0$, show that the discrete eigenvalue becomes

$$\kappa_1 \sim \tfrac{1}{2}\varepsilon^2 \int_{-\infty}^{\infty} \left\{ \int_{x}^{\infty} f(x')\,dx' \right\}^2 dx \qquad \text{as} \quad \varepsilon \to 0^+.$$

[Hints: for (i) use a straightforward asymptotic expansion; for (ii) use multiple scales e.g. $x, X = \varepsilon x$ with $\psi = A(X;\varepsilon)\exp\{\varepsilon g(x,X;\varepsilon)\}$.]

(Landau & Lifshitz, 1977; Morse & Feshbach, 1953; Miles, 1978b; Drazin, 1963.)

Q3.9 *A* sech^2 *potential.* Verify the results (3.22), for $a(k)$ and $b(k)$, by working from expression (3.21), for $\hat{\psi}$, and using the properties of the hypergeometric function.

Q3.10 *The scattering coefficients.* With $u(x) = -U_0 \operatorname{sech}^2 x$ show, by using the properties of the gamma function, that $a(k)$ and $b(k)$ (as given by equations (3.22) and also obtained in Q3.9) satisfy the conditions

(i) $|a|^2 + |b|^2 = 1$;

(ii) $a \to 1$ and $b \to 0$ as $k \to \infty$.

Q3.11 *Inverse scattering about* $-\infty$. Find the equation for $L(x, z)$ if

$$\psi_-(x; k) = e^{-ikx} + \int_{-\infty}^{x} L(x, z) e^{-ikz} \, dz$$

is a solution of $\psi'' + \{\lambda - u(x)\}\psi = 0$; see equation (3.39). What boundary conditions must $L(x, z)$ satisfy?

Q3.12 *Poles of the transmission coefficient.*

(i) Use the identity (3.43) to prove that a^{-1} has zeros at $k = i\kappa_n$.

(ii) Use the identity (3.44), et seq., to show that $a(k)$ has *simple* poles at $k = i\kappa_n$.

Q3.13 *Integral equations.* Find the solutions of the integral equations

(i) $K(x, z) + e^{-(x+z)} + \int_{x}^{\infty} K(x, y) e^{-(y+z)} \, dy = 0$;

(ii) $\phi(x, z) + xz + \int_{0}^{1} \phi(x, y) yz \, dy = 0$;

(iii) $K(x, z) - e^{-(x+z)} - \int_{-z}^{x} K(x, y) e^{-(y+z)} \, dy = 0$;

(iv) $\phi(x) = 1 + \int_{0}^{\pi} \phi(y) \sin(x + y) \, dy$.

Q3.14 *Neumann series.* Now use a Neumann series to find the solution to Q3.13(i).

Q3.15 *Inverse scattering.* Reconstruct the potential function, $u(x)$, for which the reflection coefficient is

$$b(k) = -\beta/(\beta + ik), \qquad \beta < 0.$$

Q3.16 *Inverse scattering with zero reflection coefficient.* For the case of three discrete eigenvalues, with $b(k) = 0$ for all k, find an expression for $|A|$ (see example (ii), §3.4) so that the potential function can be written as

$$u(x) = -2 \frac{d^2}{dx^2} \log |A|.$$

The initial-value problem for the Korteweg–de Vries equation

4.1 Recapitulation

In the previous chapter we discussed the classical scattering problem for the Sturm–Liouville operator, as well as the corresponding inverse scattering problem. For the purposes of the development in this chapter, where we shall describe the connection between inverse scattering and the Korteweg–de Vries equation, we need to note only the conclusions of Chap. 3. We summarise those results here.

If $u(x)$ is the potential function for the Sturm–Liouville equation

$$\psi_{xx} + (\lambda - u)\psi = 0, \qquad -\infty < x < \infty, \tag{4.1}$$

where λ is the eigenvalue, then u can be reconstructed from the scattering data. These data are described by the behaviours of the eigenfunction, ψ, in the form

$$\hat{\psi}(x; k) \sim \begin{cases} e^{-ikx} + b(k)e^{ikx} & \text{as } x \to +\infty \\ a(k)e^{-ikx} & \text{as } x \to -\infty, \end{cases} \tag{4.2}$$

for $\lambda > 0$, with $k = \lambda^{1/2}$ for the continuous spectrum, and

$$\psi_n(x) \sim c_n \exp(-\kappa_n x) \qquad \text{as } x \to +\infty, \tag{4.3}$$

for $\lambda < 0$, with $\kappa_n = (-\lambda)^{1/2}$ for each discrete eigenvalue $(n = 1, 2, \ldots, N)$. We then showed that

$$u(x) = -2\frac{\mathrm{d}}{\mathrm{d}x} K(x, x), \tag{4.4}$$

where $K(x, z)$ is the solution of the Marchenko equation

$$K(x, z) + F(x + z) + \int_x^\infty K(x, y)F(y + z)\,\mathrm{d}y = 0 \tag{4.5}$$

and F is defined by

$$F(X) = \sum_{n=1}^N c_n^2 \exp(-\kappa_n X) + \frac{1}{2\pi}\int_{-\infty}^\infty b(k)e^{ikX}\,\mathrm{d}k. \tag{4.6}$$

Hence the determination of u requires the solution of the linear integral equation for $K(x, z)$, an equation whose solution is amenable to standard techniques (as outlined in §3.4).

4.2 Inverse scattering and the KdV equation

We first write the KdV equation in the convenient form (see equation (2.2))

$$u_t - 6uu_x + u_{xxx} = 0. \tag{4.7}$$

Then one simple way of showing a connection with the Sturm–Liouville problem is to define a function $v(x, t)$ such that

$$u = v^2 + v_x. \tag{4.8}$$

(Equation (4.8) is called the *Miura transformation*.) Direct substitution of (4.8) into equation (4.7) then yields

$$2vv_t + v_{xt} - 6(v^2 + v_x)(2vv_x + v_{xx}) + 6v_xv_{xx} + 2vv_{xxx} + v_{xxxx} = 0,$$

which can be rearranged to give

$$\left(2v + \frac{\partial}{\partial x}\right)(v_t - 6v^2v_x + v_{xxx}) = 0. \tag{4.9}$$

Thus, if v is a solution of

$$v_t - 6v^2v_x + v_{xxx} = 0 \tag{4.10}$$

(a *modified KdV equation*, sometimes called an *mKdV equation*; see Q2.1), then equation (4.8) defines a solution of the KdV equation, (4.7). (Note that a solution of equation (4.7) does not imply a solution of equation (4.10) since equation (4.9) contains an additional operator.)

Now, we recognise equation (4.8) as a *Riccati equation* for v which therefore may be linearised by the substitution

$$v = \psi_x/\psi \tag{4.11}$$

for some differentiable function $\psi(x; t) \neq 0$. The fact that time (t) occurs only parametrically in equation (4.8) is accommodated in our notation for ψ by the use of the semicolon. Equation (4.8), upon the introduction of (4.11), becomes

$$\psi_{xx} - u\psi = 0 \tag{4.12}$$

which is almost the (time-independent) Sturm–Liouville equation for ψ. The connection is completed when we observe that the KdV equation is *Galilean invariant*, that is

$$u(x, t) \rightarrow \lambda + u(x + 6\lambda t, t), \qquad -\infty < \lambda < \infty,$$

leaves equation (4.7) unchanged for arbitrary (real) λ. Since the x-

dependence is unaltered under this transformation (and t plays the rôle of a parameter) we may equally replace u by $u - \lambda$. The equation for ψ now becomes

$$\psi_{xx} + (\lambda - u)\psi = 0, \tag{4.13}$$

which is the Sturm–Liouville equation with potential u and eigenvalue λ. Thus, if we are able to solve for ψ, we can then recover u from equations (4.11) and (4.8). However, the procedure is far from straightforward since equation (4.13) already involves the function u which we wish to determine. The way to avoid this dilemma is to interpret the problem in terms of scattering by the potential u.

Let $u(x,t)$ be the solution of

$$u_t - 6uu_x + u_{xxx} = 0,$$

with $u(x,0) = f(x)$ given: this defines the initial-value problem for the KdV equation. Further, let us introduce the function ψ which satisfies the equation

$$\psi_{xx} + (\lambda - u)\psi = 0,$$

for some λ, and by virtue of the parametric dependence on t we must allow $\lambda = \lambda(t)$. The solution of the KdV equation can therefore be described as follows. At $t = 0$ we are given $u(x,0) = f(x)$ and so (provided ψ exists) we may solve the scattering problem for this potential, yielding expressions for $b(k)$, κ_n and c_n $(n = 1,\ldots,N)$. If the *time evolution* of these scattering data can be determined then we shall know the scattering data at any later time. This information therefore allows us to solve the inverse scattering problem and so reconstruct $u(x,t)$ for $t > 0$. The procedure is represented schematically in Fig. 4.1, where $S(t)$ denotes the scattering data i.e. $b(k;t)$, $\kappa_n(t)$ and $c_n(t)$ $(n = 1,\ldots,N)$.

It is clear that the success, or failure, of this approach now rests on whether, or not, the time evolution of S can be determined. Furthermore, it is to be hoped that the evolution is fairly straightforward so that

Fig. 4.1 Representation of the inverse scattering transform for the KdV equation.

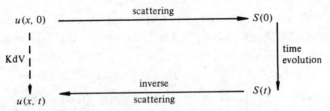

application of this technique does not prove too difficult. We shall demonstrate in the next section how $S(t)$ can be found and also show that it takes a surprisingly simple form. However, before we start this, we note the parallel between the scheme represented in Fig. 4.1 and the use of the Fourier transform for the solution of linear partial differential equations. Consider the example discussed in §1.1,

$$u_t + u_x + u_{xxx} = 0,$$

which is one linearisation of the KdV equation. If $u(x, 0) = f(x)$ then we can write

$$f(x) = \int_{-\infty}^{\infty} A(k) e^{ikx} \, dk$$

or

$$A(k) = \frac{1}{2\pi} \int_{-\infty}^{\infty} f(x) e^{-ikx} \, dx,$$

and $A(k)$ is then analogous to the scattering data $S(0)$. Further, if

$$u(x, t) = \int_{-\infty}^{\infty} A(k) e^{i(kx - \omega t)} \, dk$$

where $\omega = \omega(k)$, then

$$\omega(k) = k - k^3,$$

and the term in ω expresses the time evolution of the 'scattering data'. This procedure is shown schematically in Fig. 4.2 where the Fourier transform (F.T.), and its inverse, correspond to the scattering, and inverse scattering processes, respectively.

4.3 Time evolution of the scattering data

The first – and probably most surprising – result concerns the constancy of the discrete spectrum as $u(x, t)$ evolves in time. (Remember that the

Fig. 4.2 Representation of the use of the Fourier transform for a linear partial differential equation (pde).

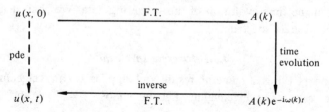

parametric dependence on t requires $\lambda = \lambda(t)$, in general.) To show this, we start with the Sturm–Liouville problem for $\psi(x;t)$ where

$$\psi_{xx} + (\lambda - u)\psi = 0, \qquad -\infty < x < \infty, \qquad (4.14)$$

and assume that this equation may be differentiated with respect to x,

$$\psi_{xxx} - u_x\psi + (\lambda - u)\psi_x = 0 \qquad (4.15)$$

and with respect to t,

$$\psi_{xxt} + (\lambda_t - u_t)\psi + (\lambda - u)\psi_t = 0, \qquad (4.16)$$

where $u(x,t)$ satisfies the KdV equation

$$u_t - 6uu_x + u_{xxx} = 0. \qquad (4.17)$$

We shall make use of these results in what follows. Now, it is convenient to define

$$R(x,t) = \psi_t + u_x\psi - 2(u + 2\lambda)\psi_x \qquad (4.18)$$

and then to construct the identity

$$\frac{\partial}{\partial x}(\psi_x R - \psi R_x) = \psi_{xx}(\psi_t + u_x\psi - 2u\psi_x - 4\lambda\psi_x)$$
$$- \psi(\psi_{xxt} + u_{xxx}\psi - 3u_x\psi_{xx} - 2u\psi_{xxx} - 4\lambda\psi_{xxx}).$$

Upon the elimination of ψ_{xxx} and ψ_{xxt} (using equations (4.15) and (4.16), respectively) we obtain

$$\frac{\partial}{\partial x}(\psi_x R - \psi R_x) = \psi_{xx}(\psi_t - 2u\psi_x - 4\lambda\psi_x) - \psi(u_{xxx}\psi - 4u_x\psi_{xx})$$
$$- \psi(u\psi_t - \lambda\psi_t - \lambda_t\psi + u_t\psi) + \psi(2u + 4\lambda)(u_x\psi - \lambda\psi_x + u\psi_x)$$

which simplifies (on using equation (4.14)) to give

$$\frac{\partial}{\partial x}(\psi_x R - \psi R_x) = \psi^2(\lambda_t - u_t + 6uu_x - u_{xxx}),$$

and since u satisfies the KdV equation, (4.17), we finally have

$$\frac{\partial}{\partial x}(\psi_x R - \psi R_x) = \lambda_t\psi^2. \qquad (4.19)$$

This equation can now be treated separately, for $\lambda < 0$ and $\lambda > 0$, in order to obtain the time evolution of the scattering data. We shall deal, first, with the discrete spectrum.

4.3.1 Discrete spectrum

Equation (4.19) is a general result and applies to any eigenfunction and corresponding eigenvalue. Let us choose $\lambda = -\kappa_n^2$ and $\psi = \psi_n$

(for $n = 1, 2, \ldots, N$), and then integrate equation (4.19) over all x,

$$[\psi_{nx} R_n - \psi_n R_{nx}]_{-\infty}^{\infty} = -(\kappa_n^2)_t \int_{-\infty}^{\infty} \psi_n^2 \, dx,$$

where R_n denotes R evaluated in terms of κ_n and ψ_n. Since the eigenfunctions, ψ_n, are normalised according to

$$\int_{-\infty}^{\infty} \psi_n^2 \, dx = 1,$$

and ψ_n – and so also R_n – decay exponentially as $|x| \to \infty$, we obtain

$$(\kappa_n^2)_t = 0 \qquad \text{or} \qquad \kappa_n = \text{constant}. \tag{4.20}$$

Thus each discrete eigenvalue $-\kappa_n^2$ $(n = 1, 2, \ldots, N)$ is a *constant of the motion*: once the κ_ns have been determined for a given initial potential (i.e. wave profile), they are fixed for all time.

If we return to equation (4.19), and introduce $\kappa_n = $ constant, then an integration with respect to x gives

$$\psi_{nx} R_n - \psi_n R_{nx} = g_n(t), \tag{4.21}$$

where g_n $(n = 1, 2, \ldots, N)$ are arbitrary functions of t. However, both ψ_n and R_n approach zero as $|x| \to \infty$ and hence, for each n, $g_n = 0$ for all t. Thus from equation (4.21), after a further integration, we obtain

$$R_n / \psi_n = h_n(t),$$

where the h_n $(n = 1, 2, \ldots, N)$ are also arbitrary functions of t. If we multiply this equation by ψ_n^2, then we obtain

$$\psi_n(\psi_{nt} + u_x \psi_n - 2u\psi_{nx} + 4\kappa_n^2 \psi_{nx}) = h_n \psi_n^2$$

or

$$\tfrac{1}{2}(\psi_n^2)_t + (u\psi_n^2 - 2\psi_{nx}^2 + 4\kappa_n^2 \psi_n^2)_x = h_n \psi_n^2, \tag{4.22}$$

where equation (4.14) has been used. Again we integrate over all x, and note that the limits are independent of t, to give

$$\frac{1}{2} \frac{d}{dt} \left(\int_{-\infty}^{\infty} \psi_n^2 \, dx \right) = h_n \int_{-\infty}^{\infty} \psi_n^2 \, dx.$$

Since ψ_n is normalised (as we mentioned earlier) this equation implies that $h_n = 0$ for each n and all t, and so

$$R_n = \psi_{nt} + u_x \psi_n - 2(u - 2\kappa_n^2)\psi_{nx} = 0, \tag{4.23}$$

which can be regarded as a time-evolution equation for $\psi_n(x; t)$. Equation (4.23) can now be used directly to find the time evolution of the normalisation 'constant', $c_n(t)$. We know that

$$u \to 0 \qquad \text{and} \qquad \psi_n(x; t) \sim c_n(t) \exp(-\kappa_n x) \qquad \text{as } x \to +\infty,$$

and this asymptotic behaviour used in equation (4.23) requires that

$$\frac{\mathrm{d}c_n}{\mathrm{d}t} - 4\kappa_n^3 c_n = 0 \qquad \text{or} \qquad c_n(t) = c_n(0)\exp(4\kappa_n^3 t), \qquad (4.24)$$

where $c_n(0)$, $n = 1, 2\ldots, N$, are the normalisation constants determined at $t = 0$.

4.3.2 Continuous spectrum

Essentially the same procedure can be adopted in this case, for which $\lambda = k^2 \, (> 0)$. Since k may take any real value, we are at liberty to follow the time evolution at k fixed. With this choice, and $\psi = \hat{\psi}$ (the continuous eigenfunction), equation (4.19) is integrated once to give

$$\hat{\psi}_x \hat{R} - \hat{\psi} \hat{R}_x = g(t; k)$$

where g is a function of integration, and $\hat{R}$ denotes R evaluated in terms of $\hat{\psi}$. For the continuous eigenfunction we have

$$\hat{\psi}(x; t, k) \sim a(k; t)\mathrm{e}^{-ikx} \qquad \text{as } x \to -\infty,$$

and so from the definition of R, (4.18), we see that

$$\hat{R}(x, t; k) \sim \left(\frac{\mathrm{d}a}{\mathrm{d}t} + 4ik^3 a\right)\mathrm{e}^{-ikx} \qquad \text{as } x \to -\infty$$

and therefore

$$\hat{\psi}_x \hat{R} - \hat{\psi} \hat{R}_x \to 0 \qquad \text{as } x \to -\infty.$$

Thus $g(t; k) = 0$ for all t, and integrating once more (paralleling the case for the discrete spectrum) we obtain

$$\hat{R}/\hat{\psi} = h(t; k) \qquad \text{or} \qquad \hat{R} = h\hat{\psi}, \qquad (4.25)$$

where h is another function of integration. If we now introduce the asymptotic behaviour of $\hat{\psi}$ and $\hat{R}$, as $x \to -\infty$, then equation (4.25) requires that

$$\frac{\mathrm{d}a}{\mathrm{d}t} + 4ik^3 a = ha. \qquad (4.26)$$

The corresponding behaviour as $x \to +\infty$ is

$$\hat{R}(x, t; k) \sim \frac{\mathrm{d}b}{\mathrm{d}t}\mathrm{e}^{ikx} + 4ik^3(\mathrm{e}^{-ikx} - b\mathrm{e}^{ikx})$$

with $\hat{\psi}(x; t, k) \sim \mathrm{e}^{-ikx} + b(k; t)\mathrm{e}^{ikx}$, and so equation (4.25) implies that

$$\frac{\mathrm{d}b}{\mathrm{d}t}\mathrm{e}^{ikx} + 4ik^3(\mathrm{e}^{-ikx} - b\mathrm{e}^{ikx}) = h(\mathrm{e}^{-ikx} + b\mathrm{e}^{ikx}). \qquad (4.27)$$

Because e^{ikx} and e^{-ikx} are linearly independent functions, equation (4.27) gives the two conditions

$$\frac{db}{dt} - 4ik^3 b = hb; \qquad h(t; k) = 4ik^3 \qquad (4.28)$$

and so equation (4.26) becomes simply

$$\frac{da}{dt} = 0. \qquad (4.29)$$

Hence, solving for a and b, we obtain

$$a(k; t) = a(k; 0); \qquad b(k; t) = b(k; 0)\exp(8ik^3 t), \qquad \text{for } t \geqslant 0, \quad (4.30)$$

which describe the time evolution of the scattering coefficients, although only the reflection coefficient (b) actually varies with t.

This completes the determination of the time evolution of the scattering data, which we summarise in the form

$$\left. \begin{array}{l} \kappa_n = \text{constant}; \quad c_n(t) = c_n(0)\exp(4\kappa_n^3 t) \\[2mm] \text{and} \\[2mm] b(k; t) = b(k; 0)\exp(8ik^3 t). \end{array} \right\} \quad (4.31\text{a, b, c})$$

Indeed, we see that the time dependence takes a particularly simple form when $u(x, t)$ evolves according to the KdV equation. (We might ask whether, if $u(x, t)$ were to evolve according to any other differential equation, the scattering data would have as simple a time dependence. We shall pursue this particular generalisation in Chaps. 5 and 6.)

4.4 Construction of the solution: summary

We now describe how the *inverse scattering transform* can be used to construct the solution to the initial-value problem for the KdV equation. In this section we shall summarise the results, and then in §§4.5 and 4.6 we discuss some specific examples.

We wish to solve the KdV equation

$$u_t - 6uu_x + u_{xxx} = 0, \qquad t > 0, \quad -\infty < x < \infty,$$

with $u(x, 0) = f(x)$. It is assumed that f is a sufficiently well-behaved function in order to ensure the existence of a solution of the KdV equation and also of the Sturm–Liouville equation

$$\psi_{xx} + (\lambda - u)\psi = 0, \qquad -\infty < x < \infty. \qquad (4.32)$$

The first stage is to set $u(x, 0) = f(x)$ and solve equation (4.32), at least to the extent of determining the discrete spectrum, $-\kappa_n^2$, the normalisation constants, $c_n(0)$, and the reflection coefficient, $b(k; 0)$. The time evolution of these scattering data is then given by equations (4.31),

$$\kappa_n = \text{constant}; \quad c_n(t) = c_n(0) \exp(4\kappa_n^3 t); \quad b(k; t) = b(k; 0) \exp(8ik^3 t).$$

The function F, defined by equation (4.6) and required for the Marchenko equation, is written as

$$F(X; t) = \sum_{n=1}^{N} c_n^2(0) \exp(8\kappa_n^3 t - \kappa_n X) + \frac{1}{2\pi} \int_{-\infty}^{\infty} b(k; 0) \exp(8ik^3 t + ikX) \, dk.$$

Note that F also depends upon the parameter t.

The Marchenko equation, (4.5), for $K(x, z; t)$, is therefore

$$K(x, z; t) + F(x + z; t) + \int_{x}^{\infty} K(x, y; t) F(y + z; t) \, dy = 0$$

(although the dependence on t here is often suppressed). Finally, the solution of the KdV equation can be expressed as

$$u(x, t) = -2 \frac{\partial}{\partial x} \hat{K}(x, t) \quad \text{and} \quad \hat{K}(x, t) = K(x, x, t),$$

where we have used equation (4.4) and again made the time dependence explicit.

This completes the presentation of the method for solving the KdV equation. It is clear, however, that we shall still be involved in two stages which could prove technically difficult, namely solving both the Sturm–Liouville equation for the given $u(x, 0) = f(x)$, and the Marchenko equation. Nevertheless, it should be emphasised that this technique has reduced the solving of a *nonlinear partial* differential equation to that of solving two *linear* problems (a second order *ordinary* differential equation, and an *ordinary* integral equation).

4.5 Reflectionless potentials

The inverse scattering transform method is best exemplified by choosing the initial profile, $u(x, 0)$, to be a sech^2 function – and in particular one of those which corresponds to a reflectionless potential (i.e. $b(k) = 0$ for all k). In this section we shall describe three such examples: first, the solitary-wave solution (sometimes called the single-soliton solution), then the two-soliton solution and finally the N-soliton solution. Although the solitary wave is already known to be an exact solution of the KdV equation (see §1.2), it is possible to obtain this solution by posing a suitable initial-value problem without the assumption that the solution takes the form of a steady progressing wave. This first example then affords a simple introduction to the application of the technique.

Example (i): solitary wave

The initial profile is taken to be

$$u(x, 0) = - 2 \operatorname{sech}^2 x \tag{4.33}$$

and so the Sturm–Liouville equation, at $t = 0$, becomes

$$\psi_{xx} + (\lambda + 2 \operatorname{sech}^2 x)\psi = 0.$$

From example (ii) in §3.2, we see that

$$\frac{d}{dT}\left\{(1 - T^2)\frac{d\psi}{dT}\right\} + \left(2 + \frac{\lambda}{1 - T^2}\right)\psi = 0$$

where $T = \tanh x$, and then the only bounded solution for $\lambda = - \kappa^2 \ (<0)$ occurs if $\kappa = \kappa_1 = 1$. The corresponding discrete eigenfunction is

$$\psi_1(x) \propto P_1^1(\tanh x) = - \operatorname{sech} x,$$

and since

$$\int_{-\infty}^{\infty} \operatorname{sech}^2 x \, dx = 2$$

the normalised eigenfunction becomes

$$\psi_1(x) = 2^{-1/2} \operatorname{sech} x.$$

(The sign of ψ_1 is irrelevant.) The asymptotic behaviour of this solution yields

$$\psi_1(x) \sim 2^{1/2} e^{-x} \qquad \text{as } x \to + \infty$$

so that $c_1(0) = 2^{1/2}$, and then equation (4.31b) gives

$$c_1(t) = 2^{1/2} e^{4t}.$$

This information is sufficient for the reconstruction of $u(x, t)$ since we have chosen an initial profile for which $b(k) = 0$ for all k (see §3.2, example (ii)).

Now, from §4.4, we obtain

$$F(X; t) = 2 e^{8t - X}$$

which incorporates only one term from the summation, the contribution from the integral being zero. The Marchenko equation is therefore

$$K(x, z; t) + 2 e^{8t - (x + z)} + 2 \int_{x}^{\infty} K(x, y; t) e^{8t - (y + z)} dy = 0,$$

which implies

$$K(x, z; t) = L(x, t) e^{-z}$$

for some function L such that

$$L + 2e^{8t-x} + 2Le^{8t} \int_x^\infty e^{-2y} \, dy = 0.$$

This can be solved directly to yield

$$L(x,t) = \frac{-2e^{8t-x}}{1 + e^{8t-2x}}$$

and then

$$u(x,t) = 2\frac{\partial}{\partial x}\left(\frac{2e^{8t-2x}}{1 + e^{8t-2x}}\right)$$

$$= -\frac{8e^{2x-8t}}{(1 + e^{2x-8t})^2}$$

$$= -2\operatorname{sech}^2(x - 4t), \tag{4.34}$$

which is the solitary wave of amplitude -2 and speed of propagation 4 (cf. equation (2.16) with $a = 4$).

Example (ii): two-soliton solution

In this example we make full use of the scattering problem described in example (ii) of §3.2 (which the reader is advised to consult). Thus we consider the problem for which the initial profile is

$$u(x,0) = -6\operatorname{sech}^2 x \tag{4.35}$$

so that the Sturm–Liouville equation, at $t = 0$, becomes

$$\psi_{xx} + (\lambda + 6\operatorname{sech}^2 x)\psi = 0$$

or

$$\frac{d}{dT}\left\{(1 - T^2)\frac{d\psi}{dT}\right\} + \left(6 + \frac{\lambda}{1 - T^2}\right)\psi = 0$$

where $T = \tanh x$. This equation has bounded solutions, for $\lambda = -\kappa^2 \, (<0)$ if $\kappa_1 = 1$ or $\kappa_2 = 2$, of the form

$$\psi_1(x) = \sqrt{\left(\frac{3}{2}\right)}\tanh x \operatorname{sech} x; \quad \psi_2(x) = \frac{\sqrt{3}}{2}\operatorname{sech}^2 x$$

both of which have been made to satisfy the normalisation condition. The asymptotic behaviours of these solutions are

$$\psi_1(x) \sim \sqrt{6}e^{-x}; \qquad \psi_2(x) \sim 2\sqrt{3}e^{-2x} \qquad \text{as } x \to +\infty$$

so that

$$c_1(0) = \sqrt{6}; \qquad c_2(0) = 2\sqrt{3};$$

and then

$$c_1(t) = \sqrt{6}e^{4t}; \qquad c_2(t) = 2\sqrt{3}e^{32t}.$$

As in example (i) above, the choice of initial profile ensures that $b(k) = 0$ for all k and so $b(k; t) = 0$ for all t. The function F then becomes

$$F(X; t) = 6e^{8t - X} + 12e^{64t - 2X}$$

(since there are two terms in the series), and the Marchenko equation is therefore

$$K(x, z; t) + 6e^{8t - (x+z)} + 12e^{64t - 2(x+z)}$$
$$+ \int_x^\infty K(x, y; t)\{6e^{8t - (y+z)} + 12e^{64t - 2(y+z)}\}\, dy = 0.$$

(This is a particular version of equation (3.55).) It is clear that the solution for K must take the form

$$K(x, z; t) = L_1(x, t)e^{-z} + L_2(x, t)e^{-2z}$$

since F is a separable function (see §3.4). Collecting the coefficients of e^{-z} and e^{-2z}, we obtain the pair of equations

$$L_1 + 6e^{8t - x} + 6e^{8t}\left(L_1 \int_x^\infty e^{-2y}\, dy + L_2 \int_x^\infty e^{-3y}\, dy\right) = 0;$$

$$L_2 + 12e^{64t - 2x} + 12e^{64t}\left(L_1 \int_x^\infty e^{-3y}\, dy + L_2 \int_x^\infty e^{-4y}\, dy\right) = 0,$$

for the functions L_1 and L_2. Upon the evaluation of the definite integrals, these two equations become

$$L_1 + 6e^{8t - x} + 3L_1 e^{8t - 2x} + 2L_2 e^{8t - 3x} = 0;$$
$$L_2 + 12e^{64t - 2x} + 4L_1 e^{64t - 3x} + 3L_2 e^{64t - 4x} = 0,$$

which are easily solved to yield

$$L_1(x, t) = 6(e^{72t - 5x} - e^{8t - x})/D; \qquad L_2(x, t) = -12(e^{64t - 2x} + e^{72t - 4x})/D$$

where $D(x, t) = 1 + 3e^{8t - 2x} + 3e^{64t - 4x} + e^{72t - 6x}$.

The solution of the KdV equation can now be expressed as

$$u(x, t) = -2\frac{\partial}{\partial x}(L_1 e^{-x} + L_2 e^{-2x})$$

$$= 12\frac{\partial}{\partial x}\{(e^{8t - 2x} + e^{72t - 6x} - 2e^{64t - 4x})/D\},$$

which can be simplified (after a little manipulation) to give

$$u(x, t) = -12\frac{3 + 4\cosh(2x - 8t) + \cosh(4x - 64t)}{\{3\cosh(x - 28t) + \cosh(3x - 36t)\}^2}, \qquad (4.36)$$

the two-soliton solution. (The details in the above calculation are left as an exercise.)

Here we digress to describe some properties of the solution, and to explain why it is a 'two-soliton' solution. To do this we shall present both numerical results obtained directly from solution (4.36), and some asymptotic results valid as $t \to \pm \infty$. Since the solution is valid for all positive and negative t, we may examine the development of the profile both before and after the formation of the initial profile, (4.35), specified at $t = 0$. The wave profile, plotted as a function of x at five different times, is shown in Fig. 4.3. (Note that we have chosen to plot $-u$ rather than u; this allows a direct comparison to be made with the application of the KdV equation to water waves that was described in §1.3.)

The solution depicts two waves (which are almost solitary) where the taller one catches the shorter, coalesces to form a single wave – our initial profile at $t = 0$ – and then reappears to the right and moves away from

Fig. 4.3 The two-soliton solution with $u(x, 0) = -6 \operatorname{sech}^2 x$ (see (c)); (a) $t = -0.5$; (b) $t = -0.1$; (d) $t = 0.1$; (e) $t = 0.5$. Note that $-u$ is plotted against x.

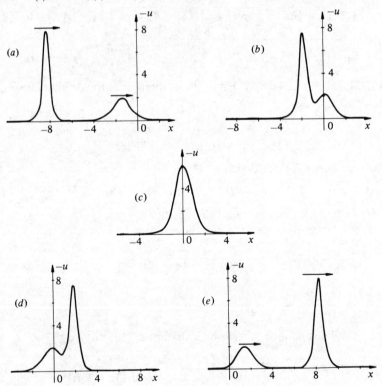

the shorter one as t increases. This interaction might seem, at first sight, to be a purely linear process but this is not so. A more careful examination of the plots shows that the taller wave has moved *forward*, and the shorter one *backward*, relative to the positions they would have reached if the interaction was indeed linear. This important observation is seen particularly clearly in Fig. 4.4 where the paths of the wave crests are represented in the (x, t)-plane. In fact it is the appearance of phase shifts which is the hallmark of this type of nonlinear interaction. Each well-defined solitary-like wave which occurs as $t \to \pm \infty$, and which interacts in this special way, is called a *soliton*; we have therefore described the two-soliton solution.

The character of this solution is also made evident by examining the asymptotic behaviour of $u(x, t)$ as $t \to \pm \infty$. For example, if we introduce $\xi = x - 16t$, then solution (4.36) can be expressed as

$$u(x, t) = -12 \frac{3 + 4 \cosh(2\xi + 24t) + \cosh(4\xi)}{\{3 \cosh(\xi - 12t) + \cosh(3\xi + 12t)\}^2}$$

which can be expanded as $t \to \pm \infty$, at ξ fixed. This asymptotic limit ensures that we follow the development of the wave which moves at a

Fig. 4.4 The paths of the two wave crests, before and after the interaction. The region where the dominant interaction occurs is inside the circle. The dotted lines show the paths of the wave crests if linear superposition were to apply.

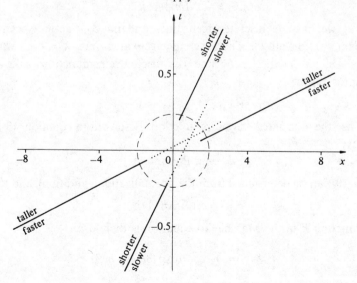

speed 16 (if such a one exists). We thus obtain

$$u(x, t) \sim -8 \operatorname{sech}^2 (2\xi \mp \tfrac{1}{2} \log 3) \qquad \text{as } t \to \pm \infty, \quad \xi = x - 16t,$$

and a similar procedure can be adopted for the wave which moves at the speed 4: let $\eta = x - 4t$, then

$$u(x, t) \sim -2 \operatorname{sech}^2 (\eta \pm \tfrac{1}{2} \log 3) \qquad \text{as } t \to \pm \infty$$

at η fixed. In fact these two asymptotic wave forms can be combined to produce a uniformly valid solution, since the error terms are exponentially small, where

$$u(x, t) \sim -8 \operatorname{sech}^2 (2\xi \mp \tfrac{1}{2} \log 3) - 2 \operatorname{sech}^2 (\eta \pm \tfrac{1}{2} \log 3) \qquad \text{as } t \to \pm \infty.$$

$$(4.37)$$

The solution is therefore comprised of two solitary waves at infinity, with the phase shifts now explicit. From solution (4.37) we see that the taller wave moves forward by an amount $x = \tfrac{1}{2} \log 3$, and the shorter moves back by $x = \log 3$. Finally, in view of the discussion in §4.6, it should be noted that the solution here contains no other component (such as, for example, an oscillatory dispersive wave).

Example (iii): N-soliton solution

The method that we have used in example (ii) above can be generalised by introducing the matrix formulation given in example (ii) of §3.4. The initial profile is now taken to be

$$u(x, 0) = -N(N + 1) \operatorname{sech}^2 x,$$

so that we have N discrete eigenvalues and no continuous spectrum (i.e. $b(k) = 0$ for all k). The discrete eigenvalues are $\lambda = -\kappa^2$ where $\kappa = \kappa_n = n$, for $n = 1, 2, \ldots, N$, and the discrete eigenfunctions take the asymptotic form

$$\psi_n(x) \sim c_n \mathrm{e}^{-nx} \qquad \text{as } x \to +\infty.$$

If we use the associated Legendre functions described in equations (3.19), we have

$$\psi_n(x) \propto P_N^n(\tanh x)$$

(and $c_n(0)$ can be determined from the normalisation condition), and then

$$c_n(t) = c_n(0) \exp(4n^3 t).$$

The function F in the Marchenko equation is therefore

$$F(X; t) = \sum_{n=1}^{N} c_n^2(0) \exp(8n^3 t - nX),$$

and so

$$K(x,z;t) + \sum_{n=1}^{N} c_n^2(0) \exp\{8n^3 t - n(x+z)\}$$

$$+ \int_x^{\infty} K(x,y;t) \sum_{n=1}^{N} c_n^2(0) \exp\{8n^3 t - n(y+z)\} \, dy = 0.$$

The solution for K must now take the form

$$K(x,z;t) = \sum_{n=1}^{N} L_n(x,t) e^{-nz},$$

and if we follow example (ii) in §3.4 then the integral equation is replaced by the algebraic system

$$AL + B = 0,$$

where L and B are column vectors with elements L_n and $B_n = c_n^2(0) \exp(8n^3 t - nx)$, respectively. The $N \times N$ matrix A has the elements

$$A_{mn} = \delta_{mn} + \frac{c_m^2(0)}{m+n} \exp\{8m^3 t - (m+n)x\},$$

and so from equation (3.58) we can write the solution as

$$u(x,t) = -2 \frac{\partial^2}{\partial x^2} \log|A|.$$

As in the previous example, the asymptotic form of the solution can be determined. Thus, if $\xi_n = x - 4\kappa_n^2 t = x - 4n^2 t$ and $t \to \pm\infty$ at ξ_n fixed, the behaviour of u is

$$u(x,t) \sim -2n^2 \operatorname{sech}^2\{n(x - 4n^2 t) \mp x_n\}$$

where the phase, x_n, is given by

$$\exp(2x_n) = \prod_{\substack{m=1 \\ m \neq n}}^{N} \left,\left|\frac{n-m}{n+m}\right|^{\operatorname{sgn}(n-m)},$$

for $n = 1, 2, \ldots, N$. Also, the N-soliton form is evident from the uniformly valid solution

$$u(x,t) \sim -2 \sum_{n=1}^{N} n^2 \operatorname{sech}^2\{n(x - 4n^2 t) \mp x_n\} \qquad \text{as } t \to \pm\infty.$$

Thus the asymptotic solution represents separate solitons, ordered according to their speeds: as $t \to +\infty$ the tallest (and therefore fastest) soliton is at the front followed by progressively shorter solitons behind. All N solitons interact at $t = 0$ to form the single sech^2 pulse which was specified as the initial profile at that instant. Some plots of the three-soliton

Fig. 4.5 The three-soliton solution with $u(x,0) = -12\,\mathrm{sech}^2 x$ (see (a));
(b) $t = 0.05$; (c) $t = 0.2$. Note that $-u$ is plotted against x.

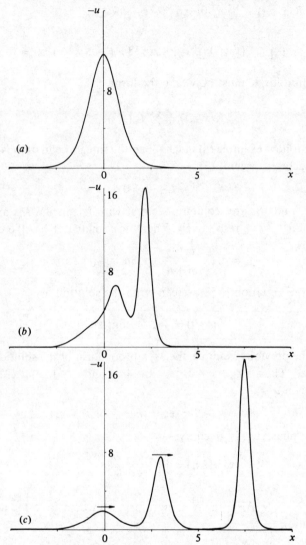

solution ($N = 3$; $u(x, 0) = -12 \operatorname{sech}^2 x$) are given in Fig. 4.5, where the emerging solitons are of amplitudes 18, 8 and 2.

4.6 Description of the solution when $b(k) \neq 0$

So far we have restricted our discussion to reflectionless initial profiles, but it is clear that a more general choice of $u(x, 0)$ will give rise to a non-zero $b(k)$. Unfortunately, when $b(k) \neq 0$, it is not possible to solve the Marchenko equation for K in closed form. We must content ourselves with a description of the solution based on numerical or asymptotic analyses, or preferably both. In this section we shall give a brief outline of the problem associated with the choice $u(x, 0) = -U_0 \delta(x)$ (see example (i) in §§3.2 and 3.4), and also we shall present numerical solutions with $u(x, 0) = -U_0 \operatorname{sech}^2 x$, for two values of U_0 which do *not* take the form $N(N + 1)$ where N is a positive integer.

Example (i): delta-function initial profile

In this example we choose

$$u(x, 0) = -U_0 \delta(x),$$

where U_0 is a constant (which we shall assume for the moment is positive), and $\delta(x)$ is Dirac's delta function. From §3.2 we know that there is a single discrete eigenvalue $\lambda = -\kappa_1^2$, where $\kappa_1 = \frac{1}{2} U_0$ ($U_0 > 0$), and

$$\psi_1(x) = \sqrt{\kappa_1} \exp(\mp \kappa_1 x), \qquad x \gtrless 0,$$

(see equation (3.12), et seq.). The continuous spectrum exists, with

$$b(k) = -U_0 / (U_0 + 2ik),$$

(see equation (3.14)), and so we have the time evolution of the scattering data as

$$c_1(t) = \sqrt{\kappa_1} \exp(4\kappa_1^3 t) \qquad \text{and} \qquad b(k; t) = \frac{-U_0 \exp(8ik^3 t)}{U_0 + 2ik}.$$

The function F can therefore be expressed as

$$F(X; t) = \kappa_1 \exp(8\kappa_1^3 t - \kappa_1 X) - \frac{U_0}{2\pi} \int_{-\infty}^{\infty} \frac{\exp(8ik^3 t + ikX)}{U_0 + 2ik} \, dk, \quad (4.38)$$

but now the Marchenko equation for K cannot be solved completely. However, it is clear that the solution for u will incorporate the single soliton associated with the discrete eigenvalue $\kappa_1 = \frac{1}{2} U_0$; that is

$$u(x, t) \sim -\frac{1}{2} U_0^2 \operatorname{sech}^2 \{\frac{1}{2} U_0 (x - U_0^2 t - x_1)\} \quad (4.39)$$

as $t \to + \infty$ with $x - U_0^2 t$ fixed. The phase shift, x_1, is given by

$$\exp(2\kappa_1 x_1) = c_1^2(0)/2\kappa_1 \qquad \text{or} \qquad x_1 = - U_0^{-1} \log 2.$$

In other words, the asymptotic solution (4.39) is generated by the contribution to F from the first term in expression (4.38); the integral term is zero where solution (4.39) is valid.

The rôle of the integral in F can be examined by considering the limit as $t \to + \infty$ for $x < 0$; in this region of (x, t)-space the soliton is exponentially small. Although the calculation is beyond the scope of this text, we note that the integral term in expression (4.38) may be written as

$$- \frac{U_0}{2\pi} \int_{-\infty}^{\infty} \frac{\exp\{i(X^3/t)^{1/2}(8\lambda^3 + \lambda)\}}{U_0(t/X)^{1/2} + 2i\lambda} \, d\lambda$$

where $k = \lambda(X/t)^{1/2}$. This transformation indicates that the solution should be conveniently expressed in terms of the similarity variable $Xt^{-1/3}$, and t, as $t \to + \infty$. This is the case: the solution turns out to be an oscillatory dispersive wave which propagates to the left with an amplitude which decays like $t^{-1/3}$ as $t \to + \infty$. (The relationship of this wave to the oscillatory part of the Airy function is seen to be quite close, particularly when it is noted that, since the wave is of vanishingly small amplitude, it must satisfy

$$u_t + u_{xxx} = 0,$$

approximately: see equations (1.4) and (1.8).) The solution is sketched in Fig. 4.6 where again we have plotted $-u$ in order to represent a soliton of elevation.

Fig. 4.6 The delta-function initial profile (see (a)). The solution at a later time is sketched in (b). Note that $-u$ is plotted against x.

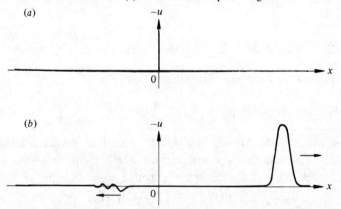

Finally, if the delta function is of positive amplitude (i.e. if $U_0 < 0$) then no discrete eigenvalue exists (see §3.2, example (i)). There is no soliton, and thus only the dispersive-wave component arises in the solution for $t > 0$: the delta function collapses, and develops into a dispersive wave train. Since this solution corresponds to an initial wave-profile with a *negative* amplitude in the context of water waves, the result agrees with Russell's observations (see §1.2, and example (iii) below).

Example (ii): a negative sech² *initial profile*

This example, and the next one, will be described solely in terms of some numerical results obtained by integrating the KdV equation itself. (For numerical methods see Chap. 7.) Here we choose

$$u(x, 0) = -4 \operatorname{sech}^2 x,$$

and from §3.2, example (ii), we have that the total number of discrete eigenvalues is

$$[(U_0 + \tfrac{1}{4})^{1/2} - \tfrac{1}{2}] + 1,$$

where [ ] denotes the integral part. Thus, with $U_0 = 4$, we shall have two discrete eigenvalues and therefore a solution with two solitons. But, since $U_0 = 4$ can not be expressed as $N(N + 1)$, for integral N, the solution will also include a dispersive-wave component. This solution is depicted in Fig. 4.7 at three different times; the appearance of two solitons is shown in the last graph. More generally, it can be shown (Segur, 1973) that an upper bound on the number of solitons, N, is given by

$$N \leqslant 1 + \tfrac{1}{2} \int_{-\infty}^{\infty} |x| \{1 + \operatorname{sgn} u(x, 0)\} u(x, 0) \, \mathrm{d}x.$$

Example (iii): a positive sech² *initial profile*

Whenever $u(x, 0) > 0$ the solution of the KdV equation will develop without the emergence of a soliton. The initial pulse will collapse and degenerate into a wave train which disperses into $x < 0$. In this example we take

$$u(x, 0) = \operatorname{sech}^2 x,$$

and the corresponding solution is given in Fig. 4.8. The relationship to the Airy function is particularly striking in the final graph which, upon turning the graph upside down (i.e. with u against $-x$), closely resembles the Ai function (see Abramowitz & Stegun, 1964).

We conclude this chapter by returning to the question of how to define a soliton. (The reader will remember that, in §1.3, we attempted to describe the properties which would distinguish a soliton from other wave-like phenomena.) Now that we have shown the connection between the KdV equation and a scattering problem, we present a less vague definition. In particular, since the solitons (as described in §1.3) are associated with the constant discrete eigenvalues of the system – indeed there is a one-to-one correspondence between them – we might propose the following definition:

Fig. 4.7 Solution with two solitons and a dispersive wave, where $u(x, 0) = -4\,\text{sech}^2 x$ (see (a)); (b) $t = 0.4$; (c) $t = 1.0$. Note that $-u$ is plotted against x.

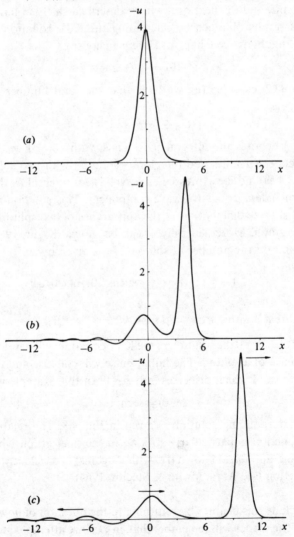

A soliton is that component of the solution of a nonlinear evolution equation which depends only upon one constant discrete eigenvalue (of the underlying scattering problem) as $t \to \pm \infty$.

One interesting aspect of this proposal is that it clarifies what is meant by the 'identity' of the soliton (see §1.3): it is that property which maintains the constancy of the discrete eigenvalues. Indeed, we can use the value of the appropriate discrete eigenvalue itself as the identifier. However, we can see that the permanent and localised natures of the wave at infinity

Fig. 4.8 Solution with dispersive wave only, where $u(x,0) = \mathrm{sech}^2 x$ (see (a)); (b) $t = 0.1$; (c) $t = 0.5$. Note that $-u$ is plotted against x.

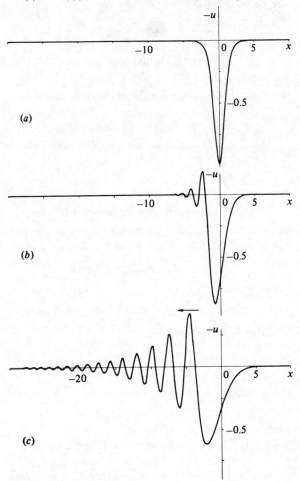

are no longer integral parts of the definition. In later chapters, where some generalisations of the inverse scattering method will be introduced, the reader may test the usefulness of our definition for the solutions of other evolution equations.

Further reading

Other accounts of the connection between the KdV equation and inverse scattering are by Miura (1976); Ablowitz & Segur (1981, Chap. 2); Dodd, Eilbeck, Gibbon & Morris, (1982, Chap. 3); Newell (1985, Chap. 1). The long-time behaviour of the solutions is discussed by Ablowitz & Segur (1981, §1.7).

Much of this work on the KdV equation was initiated by the publication of the seminal papers of Gardner, Greene, Kruskal & Miura (1967, 1974).

Exercises

*Q4.1 *Alternative derivation.*

(i) Show that, if $K(x, z)$ satisfies the Marchenko equation

$$K(x, z) + F(x, z) + \int_x^\infty K(x, y)F(y, z)\,dy = 0,$$

where F is a solution of

$$F_{xx} - F_{zz} = 0,$$

then

$$K_{xx} - K_{zz} - uK = 0, \tag{1}$$

where $u(x) = -2\,(d/dx)K(x, x)$ and K, $K_z \to 0$ as $z \to +\infty$.

(ii) Now suppose that $F = F(x, z; t)$ and $K = K(x, z; t)$, with

$$F_t + 4(F_{xxx} + F_{zzz}) = 0,$$

and show that

$$K_t + 4(K_{xxx} + K_{zzz}) - 3u_x K - 6uK_x = 0. \tag{2}$$

(iii) Use (1) and (2) to show that

$$K_t + \left(\frac{\partial}{\partial x} + \frac{\partial}{\partial z}\right)^3 K - 3u(K_x + K_z) = 0$$

and hence that

$$u_t - 6uu_x + u_{xxx} = 0.$$

*Q4.2 *Concentric KdV equation.*

(i) Show that, if $K(x, z; t)$ satisfies the Marchenko equation

$$K(x, z; t) + F(x, z; t) + \int_x^\infty K(x, y; t)F(y, z; t)\,dy = 0,$$

where F is a solution of the pair of equations

$$F_{xx} - F_{zz} = (x - z)F,$$
$$3tF_t - F + F_{xxx} + F_{zzz} = xF_x + zF_z,$$

then

$$u_t + \frac{u}{2t} - 6uu_x + u_{xxx} = 0,$$

where

$$u(x, t) = -\frac{2}{(12t)^{2/3}} \frac{\partial}{\partial X} K(X, X; t) \qquad \text{with } X = \frac{x}{(12t)^{1/3}}.$$

(ii) Hence show that a solution for F is

$$F(x, z; t) = \int_{-\infty}^{\infty} f(yt^{1/3})\text{Ai}(x + y)\text{Ai}(y + z)\,dy,$$

where f is an arbitrary function and Ai is the Airy function.

Q4.3 *Two-soliton solution of the KdV equation.* Obtain a solution for $F(x, z; t)$ (see Q4.1) which depends upon $x + z$ (but not $x - z$ since this would become trivial on $z = x$), and which is exponential in both $x + z$ and t (cf. examples (i) and (ii), §4.5).

Hence write down a solution for F which is the sum of two exponential terms, and construct the two-soliton solution of the KdV equation.

Q4.4 *Some initial-value problems.* Use the inverse scattering transform to find the solution of the KdV equation

$$u_t - 6uu_x + u_{xxx} = 0$$

which satisfies $u(x, 0) = f(x)$, $-\infty < x < \infty$, where

(i) $f(x) = -\frac{9}{2}\text{sech}^2(\frac{3}{2}x)$;

(ii) $f(x) = -12\,\text{sech}^2 x$;

*(iii) $f(x) = \begin{cases} -V & \text{for } -1 < x < 1, \\ 0 & \text{otherwise,} \end{cases}$

where $V > 0$ is a constant.

[Case (iii) is too difficult to solve explicitly, so just give a qualitative description of the solution for various V.]

Q4.5 *The character of two-soliton solutions.* Show that a special case of the two-soliton solution obtained in Q4.3 gives a sech^2 pulse at $t = 0$ (see example (ii), §4.5). Show also that, for suitably defined x, the pulse at $t = 0$ may have either one or two local maxima.

[You will find it convenient to define x so that a symmetric profile occurs at $t = 0$.]

(Lax, 1968)

Q4.6 *Three-soliton solution.* Find the asymptotic form of the three-soliton solution (see Q4.4, (ii)) as $t \to \pm\infty$, and hence determine the phase shifts.

Q4.7 *Connection with Fourier transforms.* Consider the initial-value problem for the KdV equation, with $u(x,0) = \alpha f(x)$, $-\infty < x < \infty$. Use the inverse scattering transform to show that, as $\alpha \to 0$, the first approximation to the solution for u corresponds to the solution obtained by using the Fourier transform to solve the linear problem

$$v_t + v_{xxx} = 0; \qquad v(x,0) = \alpha f(x), \qquad -\infty < x < \infty.$$

Q4.8 '*Centre of mass*' *of a solution.* Show that, if u satisfies the KdV equation, then

$$\frac{\mathrm{d}}{\mathrm{d}t} \int_{-\infty}^{\infty} xu\,\mathrm{d}x = \text{constant}, \tag{1}$$

provided $u \to 0$ (sufficiently rapidly) as $|x| \to \infty$. Interpret this result as the conservation of linear momentum of a linear mass distribution with density u. Show that (1) is consistent with the phase shifts associated with the two-soliton solution (see example (ii), §4.5).

*Q4.9 *Eigenfunction expansion* (i). Assume that u is a reflectionless potential, so that $b(k) = 0$ for all k. From example (iii), §4.5, we know that, for the N-soliton solution, we may write

$$K(x,z;t) = \sum_{n=1}^{N} L_n(x,t)\mathrm{e}^{-nz}.$$

Deduce that L_n is a solution of the Sturm–Liouville equation with $\lambda = -\kappa_n^2$, $\kappa_n = n$, and hence (or otherwise) show that

$$L_n(x,t) = -c_n(t)\psi_n(x;t),$$

where ψ_n is the discrete eigenfunction (normalised according to equation (3.5)).

Show also that

$$u(x,t) = 2 \sum_{n=1}^{N} c_n(t) \{\psi_{nx}(x;t) - \kappa_n\psi_n(x;t)\} \exp(-\kappa_n x),$$

and by using the equation satisfied by L_n (and therefore by ψ_n), deduce that

$$u(x,t) = -4 \sum_{n=1}^{N} \kappa_n \psi_n^2(x;t).$$

Hence show that, if $u(x,0)$ is a reflectionless potential and u satisfies the KdV equation, then $u(x,t) < 0$ for all $t > 0$ and for all x.

*Q4.10 *Eigenfunction expansion* (ii). Describe the form that $u(x,t)$ takes, in terms of the squared eigenfunctions, if $u(x,0)$ is no longer a reflectionless potential (see Q4.9).

Further properties of the Korteweg–de Vries equation

In this chapter we shall examine four important aspects of the KdV equation, and each one will enable us to introduce corresponding properties of other evolution equations. We shall present the ideas, therefore, primarily as they are relevant to the KdV equation; then, by use of examples and exercises, other applications will be touched upon afterwards. In this way we shall discuss: the infinity of conservation laws; the Lax formulation (in terms of operators); the Hirota bilinear form (which is particularly useful for the construction of soliton solutions); and the Bäcklund transformation between different solutions of the same equation.

Furthermore, we shall mention a few connections between these ideas. Thus we shall relate – albeit briefly – the infinity of conservation laws to a Hamiltonian structure which, in turn, is associated with a hierarchy of KdV equations. It will also be shown that the bilinear form generates a different hierarchy (but one which takes a simple form in terms of bilinear operators). Finally, we shall indicate how the bilinear form itself implies the Bäcklund transformation for the KdV equation.

It can not be emphasised enough that all the ideas developed in this chapter are generally applicable to a vast range of evolution equations which are integrable by the inverse scattering transform.

5.1 Conservation laws

5.1.1 Introduction

Conservation laws are a common feature of mathematical physics, where they describe the conservation of fundamental physical quantities. Let us consider, as an example, the familiar problem of one-dimensional gas-dynamics. If $\rho(x, t)$ is the density of the gas, and $u(x, t)$ the component of velocity in the x-direction, then the well-known equation of mass conservation is

$$\frac{\partial \rho}{\partial t} + \frac{\partial}{\partial x}(\rho u) = 0.$$

Now let us suppose that $\rho u \rightarrow$ constant as $|x| \rightarrow \infty$ (and therefore there is an ambient state). Then, if ρ and $(\rho u)_x$ are integrable for $x \in (-\infty, \infty)$, we obtain

$$\frac{d}{dt}\left(\int_{-\infty}^{\infty} \rho \, dx\right) = -[\rho u]_{-\infty}^{\infty} = 0$$

and so

$$\int_{-\infty}^{\infty} \rho \, dx = \text{constant.} \tag{5.1}$$

Equation (5.1) represents the conservation of total mass in the system, however ρ evolves in time. This simple idea can be generalised in the following manner.

First, an equation of the form

$$\frac{\partial T}{\partial t} + \frac{\partial X}{\partial x} = 0, \tag{5.2}$$

where usually neither T (the *density*) nor X (the *flux*) involve derivatives with respect to t, is called a *conservation law*. In particular, if we are to apply these ideas to an evolution equation for $u(x, t)$, then T and X may depend upon $x, t, u, u_x, u_{xx}, \ldots$, but not u_t. Now if both T and X_x are integrable on $(-\infty, \infty)$, so that

$$X \rightarrow \text{constant as } |x| \rightarrow \infty,$$

then equation (5.2) can be integrated to yield

$$\frac{d}{dt}\left(\int_{-\infty}^{\infty} T \, dx\right) = 0 \qquad \text{or} \qquad \int_{-\infty}^{\infty} T \, dx = \text{constant.} \tag{5.3}$$

The integral of T, over all x, is therefore usually called a *constant of the motion* (if we interpret t as a time-like variable).

The evolution equation of interest to us here is the KdV equation

$$u_t - 6uu_x + u_{xxx} = 0, \tag{5.4}$$

which is already in conservation form with

$$T = u \qquad \text{and} \qquad X = u_{xx} - 3u^2. \tag{5.5}$$

Thus if T and X_x are integrable, and u satisfies equation (5.4), then

$$\int_{-\infty}^{\infty} u \, dx = \text{constant,} \tag{5.6}$$

and we can see that this condition must therefore apply to all the solutions discussed in §§4.5 and 4.6 (where $u \rightarrow 0$ as $x \rightarrow \pm \infty$). Note, however, that equation (5.6) does not hold for all solutions of the KdV equation: if $u(x, t)$

is a periodic solution (see §2.4), then the analogue of condition (5.6) is the integral over just one period of the wave.

Now the KdV equation has another simple conservation law, which becomes clear if we multiply equation (5.4) by u to give

$$\frac{\partial}{\partial t}(\tfrac{1}{2}u^2) + \frac{\partial}{\partial x}(uu_{xx} - \tfrac{1}{2}u_x^2 - 2u^3) = 0.$$

Thus

$$\int_{-\infty}^{\infty} u^2 \, dx = \text{constant} \tag{5.7}$$

for all solutions, u, of the KdV equation, which vanish fast enough at infinity.

That both u and u^2 are *conserved densities* for the motion, which evolves according to the KdV equation, is to be expected. From §1.2 we know that the KdV equation describes a certain class of water waves; in particular, we have shown that u is proportional to the depth of water (relative to the ambient level) which, in turn, is also proportional to the velocity component in the x-direction. (Remember that the foregoing statements apply only in a shallow-water approximation.) Thus equation (5.6) describes the conservation of mass, and equation (5.7) the conservation of (horizontal) momentum, for the water waves. This immediately suggests that there should be a corresponding conserved density which can be associated with the *energy* of the water waves. To confirm this we construct

$$\frac{\partial}{\partial t}(u^3 + \tfrac{1}{2}u_x^2)$$

by adding $3u^2 \times (\text{KdV})$ to $u_x \times (\partial/\partial x)(\text{KdV})$. This gives

$$3u^2(u_t - 6uu_x + u_{xxx}) + u_x(u_{xt} - 6u_x^2 - 6uu_{xx} + u_{xxxx}) = 0,$$

which can be rewritten as

$$\frac{\partial}{\partial t}(u^3 + \tfrac{1}{2}u_x^2) + \frac{\partial}{\partial x}(-\tfrac{9}{2}u^4 + 3u^2u_{xx} - 6uu_x^2 + u_xu_{xxx} - \tfrac{1}{2}u_{xx}^2) = 0.$$

Since this is also a conservation law, we have a third constant of the motion, as we anticipated,

$$\int_{-\infty}^{\infty} (u^3 + \tfrac{1}{2}u_x^2) \, dx = \text{constant}. \tag{5.8}$$

At this stage one might be forgiven for believing that no further conservation laws exist. We have, after all, obtained the three standard conserved quantities which describe a one-dimensional physical system under the action of conservative forces: the conservation of mass, momentum and energy. However, by using very laborious methods, Miura, Gardner & Kruskal (1968) found *eight* more conservation laws for the KdV equation! The next two, for example, have the conserved densities

$$T_4 = 5u^4 + 10uu_x^2 + u_{xx}^2,$$
$$T_5 = 21u^5 + 105u^2u_x^2 + 21uu_{xx}^2 + u_{xxx}^2,$$

where it is obviously convenient to number them so that $T_1 = u$, $T_2 = u^2$, $T_3 = u^3 + \frac{1}{2}u_x^2$. (One of the eight – a different type – is discussed in Q5.2.) Now eleven conservation laws are more than we can reasonably expect in general, and once the three physically important ones have been exceeded you may wonder whether there is no limit. In other words, is eleven merely symptomatic of an infinity of conservation laws? It is this which we shall now investigate.

*5.1.2 An infinity of conservation laws

In the paper cited above, where the higher conservation laws were first described, an important extension was introduced; we shall follow a similar path here. First, we recall the Miura transformation, equation (4.8),

$$u = v^2 + v_x, \tag{5.9}$$

which led to the connection between the KdV equation for u and the Sturm–Liouville problem for ψ. It is now, however, convenient to work with w, rather than v, where

$$v = \tfrac{1}{2}\varepsilon^{-1} + \varepsilon w$$

and ε is an arbitrary real parameter. The Miura transformation becomes

$$u = \tfrac{1}{4}\varepsilon^{-2} + w + \varepsilon^2 w + \varepsilon w_x$$

and, since we can always shift u by an arbitrary constant (the Galilean invariance), we may just as well write

$$u = w + \varepsilon w_x + \varepsilon^2 w^2, \tag{5.10}$$

the *Gardner transformation*. (We note that w exists since v does: see §4.2.)

The Gardner transformation is now used in the KdV equation, so that

$$u_t - 6uu_x + u_{xxx} = w_t + \varepsilon w_{xt} + 2\varepsilon^2 ww_t - 6(w + \varepsilon w_x + \varepsilon^2 w^2)(w_x + \varepsilon w_{xx} + 2\varepsilon^2 ww_x)$$
$$+ w_{xxx} + \varepsilon w_{xxxx} + 2\varepsilon^2 (ww_x)_{xx}$$
$$= \left(1 + \varepsilon \frac{\partial}{\partial x} + 2\varepsilon^2 w \right)\{w_t - 6(w + \varepsilon^2 w^2)w_x + w_{xxx}\}.$$

Hence u, given by equation (5.10), is a solution of the KdV equation if w is a solution of

$$w_t - 6(w + \varepsilon^2 w^2)w_x + w_{xxx} = 0, \tag{5.11}$$

but of course not necessarily vice versa (cf. equation (4.10)). It is clear that if we set $\varepsilon = 0$ then (5.11) becomes the KdV equation and, correspondingly, the Gardner transformation reduces to $u = w$. Now equation (5.11), for all ε, is already in conservation form

$$\frac{\partial}{\partial t}(w) + \frac{\partial}{\partial x}(w_{xx} - 3w^2 - 2\varepsilon^2 w^3) = 0,$$

and so

$$\int_{-\infty}^{\infty} w \, dx = \text{constant.} \tag{5.12}$$

In order to generate conservation laws for the KdV equation, we take advantage of the arbitrary parameter ε. Since $w \to u$ as $\varepsilon \to 0$, we choose to represent w by an asymptotic expansion in ε,

$$w(x, t; \varepsilon) \sim \sum_{n=0}^{\infty} \varepsilon^n w_n(x, t) \qquad \text{as } \varepsilon \to 0. \tag{5.13}$$

(There is no requirement, therefore, that the expansion for w be convergent for some ε.) If we treat the constant in equation (5.12) similarly as a power series in ε, then, by writing w as its asymptotic expansion, we obtain

$$\int_{-\infty}^{\infty} w_n \, dx = \text{constant} \tag{5.14}$$

for each $n = 0, 1, 2, \ldots$. Finally we use the asymptotic expansion, (5.13), in the Gardner transformation, (5.10), and equate coefficients of ε^n for each $n = 0, 1, 2, \ldots$ (where u is not a function of ε). Thus from

$$\sum_{n=0}^{\infty} \varepsilon^n w_n \sim u - \varepsilon \sum_{n=0}^{\infty} \varepsilon^n w_{nx} - \varepsilon^2 \left(\sum_{n=0}^{\infty} \varepsilon^n w_n \right)^2,$$

it is a simple exercise to see that

$$w_0 = u; \qquad w_1 = -w_{0x} = -u_x; \qquad w_2 = -w_{1x} - w_0^2 = u_{xx} - u^2;$$

$$w_3 = -w_{2x} - 2w_0 w_1 = -(u_{xx} - u^2)_x + 2uu_x; \tag{5.15}$$

$$w_4 = -w_{3x} - 2w_0 w_2 - w_1^2$$

$$= -\{2uu_x - (u_{xx} - u^2)_x\}_x - 2u(u_{xx} - u^2) - u_x^2,$$

and so on. In particular we note that w_1 and w_3 are exact differentials (in x) and so the corresponding integrals, (5.14), will give us no useful information. (We assume, of course, that u is sufficiently well-behaved to ensure that each w_n is integrable on $(-\infty, \infty)$.) On the other hand, the integrals which are generated by w_0, w_2 and w_4 become our first three integrals (5.6), (5.7) and (5.8), respectively.

The results derived so far suggest that w_n is an exact differential if n is an *odd* integer and does not, therefore, lead to a conservation law. However, it would seem that if n is *even* then w_n does produce a non-trivial constant of the motion. In order to confirm this view we must show that every w_n, n odd, is an exact differential, and that, for n even, w_n is *never* an exact differential.

One of the neatest ways to achieve this is to separate the even and odd terms by associating them with the real and imaginary parts, respectively, of a new w. First, let us replace ε by a purely imaginary parameter $i\varepsilon$ (i.e. $\varepsilon \rightarrow i\varepsilon$) and then, for real u (independent of ε), $w = W$ is now a complex-valued function. Thus we have an alternative form of the Gardner transformation (5.10),

$$u = W + i\varepsilon W_x - \varepsilon^2 W^2, \tag{5.16}$$

with $W = \alpha + i\beta$ where α and β are real functions. Transformation (5.16) can therefore be written as

$$u = \alpha - \varepsilon\beta_x - \varepsilon^2(\alpha^2 - \beta^2) + i(\beta + \varepsilon\alpha_x - 2\varepsilon^2\alpha\beta), \tag{5.17}$$

and so we must have

$$\beta + \varepsilon\alpha_x - 2\varepsilon^2\alpha\beta = 0 \tag{5.18}$$

and

$$\alpha = u + \varepsilon\beta_x + \varepsilon^2(\alpha^2 - \beta^2). \tag{5.19}$$

It is convenient to write

$$\alpha \sim A_e + A_o, \qquad \beta \sim B_e + B_o, \qquad \text{as } \varepsilon \rightarrow 0,$$

where $(\ )_e$ denotes an asymptotic expansion in even powers of ε, and $(\ )_o$ is correspondingly in odd powers. If we collect all the even-power terms in equation (5.18), and all the odd-power terms in equation (5.19), then

$$\left.\begin{array}{l} B_e(1 - 2\varepsilon^2 A_e) \sim -\varepsilon(A_o)_x + 2\varepsilon^2 A_o B_o \\[2mm] A_o(1 - 2\varepsilon^2 A_e) \sim \varepsilon(B_e)_x - 2\varepsilon^2 B_e B_o, \end{array}\right\} \tag{5.20}$$

and

respectively, to all orders in ε. Since these equations imply that

$$B_e \sim -\varepsilon(A_o)_x \sim -\varepsilon^2(B_e)_{xx}, \qquad \text{as } \varepsilon \rightarrow 0,$$

we see (by iteration) that $B_e = A_o = 0$, an exact solution of equations (5.20). Thus $\alpha \sim A_e$ (even only) and $\beta \sim B_o$ (odd only) which is consistent with the result obtained by formally replacing ε by $i\varepsilon$ in the asymptotic expansion (5.13). However, we now observe that equation (5.18) can be rewritten as

$$\beta = -\varepsilon\alpha_x/(1 - 2\varepsilon^2\alpha) = \frac{1}{2\varepsilon}\frac{\partial}{\partial x}(\log|1 - 2\varepsilon^2\alpha|) \tag{5.21}$$

and so β is an exact differential to all orders in ε: the odd terms $w_1, w_3, \ldots,$ do not generate conservation laws.

Furthermore, the first iterate for α (from equation (5.19)) is

$$(\alpha)_0 = u,$$

corresponding to the term in ε^0. Thereafter the iterates will, in general, involve terms from both $\varepsilon\beta_x$ and $\varepsilon^2\beta^2$ which perforce must always include derivatives of u (see (5.21)), together with a contribution from the term $\varepsilon^2\alpha^2$. This latter term will generate, at each iteration, terms which do not involve any derivatives (together with other terms which do). These specific polynomial terms are obtained by the iteration

$$(\alpha)_m = u + \varepsilon^2\{(\alpha)_{m-1}\}^2, \qquad m = 1, 2, \ldots, \qquad (\alpha)_0 = u,$$

and so we have the sequence: $u,\ u + \varepsilon^2 u^2,\ u + \varepsilon^2(u + \varepsilon^2 u^2)^2, \ldots$. In other words, for every ε^{2n}, there will be a term proportional to u^{n+1} and such terms are evidently not exact differentials. Thus there will be an infinity of conservation laws, each density of which is characterised by the inclusion of a term in u^{n+1} (for $n = 0, 1, 2, \ldots$). (If we use our notation T_n, then each T_n includes a term u^n; see §5.1.1.)

*5.1.3 Of Lagrangians and Hamiltonians

Before we leave the conservation laws, it is worth finally – but very briefly – noting some deeper results. It is well-known that conservation laws for many systems arise from variational principles that are invariant under transformations which belong to a continuous group. For example, translations (in both x and t), Galilean invariance and rotations are such transformations. A conservation law then follows from the application of *Noether's theorem* to the *Lagrangian function* (or *density*).

Although a detailed study of Noether's theorem is beyond the scope of this text, it might be useful to digress here and give a brief outline of the essential idea. Noether considered the invariance of the *action integral* (the integral of the Lagrangian function over some time interval) to a group of infinitesimal transformations. For example, let us consider the Lagrangian $\mathscr{L}(q_i, \dot{q}_i)$, where $q_i(t)$ for $i = 1, 2, \ldots, n$ are the generalised coordinates, so that the action integral is

$$A = \int_{t_1}^{t_2} \mathscr{L}(q_i, \dot{q}_i)\, \mathrm{d}t.$$

(Note that $\mathscr{L}$ does not depend explicitly upon t.) Now $\mathscr{L}$ is invariant to the transformation $t \to t + \tau$, where τ is an arbitrary constant, and the action integral therefore transforms according to

$$A \to \int_{t_1+\tau}^{t_2+\tau} \mathscr{L}(q_i, \dot{q}_i) \, dt.$$

The Euler–Lagrange equations are thus unaltered by this transformation since τ appears only in the limits.

Noether's approach was to regard $\tau = \tau(t)$, where τ is now an infinitesimal, and such that $\tau(t_1) = \tau(t_2) = 0$. Since $\dot{q}_i \to \dot{q}_i/(1 + \dot{\tau}) \sim \dot{q}_i(1 - \dot{\tau})$ as $\tau \to 0$, we obtain

$$\mathscr{L}(q_i, \dot{q}_i) \to \mathscr{L}(q_i, \dot{q}_i) - \left(\sum_{i=1}^{n} \frac{\partial \mathscr{L}}{\partial \dot{q}_i} \dot{q}_i \right) \dot{\tau}$$

(retaining the first order in τ), and then the action integral transforms as

$$A \to \int_{t_1}^{t_2} \mathscr{L}(q_i, \dot{q}_i) \, dt - \int_{t_1}^{t_2} \left(\sum_{i=1}^{n} p_i \dot{q}_i - \mathscr{L} \right) \dot{\tau} \, dt,$$

to the same order in τ, where $p_i = \partial \mathscr{L}/\partial \dot{q}_i$ is the generalised momentum. (Remember that $dt \to (1 + \dot{\tau}) \, dt$.) This transformed variational problem now has an additional degree of freedom by virtue of $\dot{\tau}$, and the corresponding Euler–Lagrange equation is

$$\frac{d}{dt} \left(\sum_{i=1}^{n} p_i \dot{q}_i - \mathscr{L} \right) = 0$$

or

$$\sum_{i=1}^{n} p_i \dot{q}_i - \mathscr{L} = \text{constant},$$

which is the equation for the conservation of energy.

For the KdV equation

$$u_t - 6uu_x + u_{xxx} = 0$$

we set $u = \phi_x$ so that

$$\phi_{xt} - 6\phi_x \phi_{xx} + \phi_{xxxx} = 0. \tag{5.22}$$

It is then easy to confirm that the Lagrangian

$$\mathscr{L} = \tfrac{1}{2} \phi_x \phi_t - \phi_x^3 - \tfrac{1}{2} \phi_{xx}^2$$

generates this alternative form of the KdV equation, (5.22), as the corresponding Euler–Lagrange equation. Now, the invariance of $\mathscr{L}$ under translations of ϕ, x and t, together with application of Noether's theorem, reproduces the three conservation laws of mass, momentum and energy, respectively. (Since $\mathscr{L}$ is also invariant under a Galilean transformation, Noether's theorem in this case gives the conservation law discussed in Q5.2.)

Furthermore, the third conserved density (namely energy) for the KdV equation can be used directly in a *Hamiltonian* formulation of this equation by writing

$$\frac{\partial u}{\partial t} = \frac{\partial}{\partial x}\left\{\frac{\delta}{\delta u}(u^3 + \tfrac{1}{2}u_x^2)\right\},$$

where $\delta/\delta u$ is the *variational* (or *Fréchet*) derivative (see Q5.8). Indeed, by virtue of the infinity of conserved densities, the KdV equation corresponds to an *infinite Hamiltonian system*. The interpretation of an evolution equation (which can be solved by the inverse scattering transform) as a Hamiltonian system has, in recent years, become an important aspect of this work. (References to the above material will be found in the section on 'Further reading' at the end of this chapter.)

5.2 Lax formulation and its KdV hierarchy

We asked, at the end of §4.3, if the KdV equation was the only equation with all the special properties we have described here. For example, if $u(x,t)$ satisfies any other evolution equation, will the time evolution of the scattering data be as simple? One very important aspect, which goes a long way to answer this question, was developed by Lax in 1968. This work – perhaps even more than that on conservation laws – introduces far deeper and more fundamental ideas than we have met hitherto. It will soon become clear that, indeed, other equations with similar properties do exist: the KdV equation does not stand alone in this class of evolution equations.

5.2.1 Description of the method: operators

The argument developed by Lax requires the use of a little functional analysis, notably the ideas of inner product and self-adjoint operators in a Hilbert space. We shall remind the reader of the various points as we need them. This initial phase of the work is presented without reference to the KdV equation; the application to the KdV equation will be given in the next section.

First, suppose that we wish to solve the initial-value problem for u, where $u(x,t)$ satisfies some nonlinear evolution equation of the form

$$u_t = N(u), \tag{5.23}$$

with $u(x,0) = f(x)$. We assume that $u \in Y$ for all t, and that $N: Y \to Y$ is some nonlinear operator which is independent of t but may involve x or derivatives with respect to x, where Y is some appropriate function space.

It is clear, therefore, that Y need not be the space of smooth scalar functions of x, vanishing as $x \to \pm \infty$. Also N need not be a partial differential operator, although such a choice is possible, and so equation (5.23) could be the KdV equation with

$$N(u) = 6uu_x - u_{xxx}, \qquad -\infty < x < \infty.$$

Next, we suppose that the evolution equation, (5.23), can be expressed in the form

$$L_t = ML - LM \tag{5.24}$$

where L and M are some *linear* operators in x, which operate on elements of a Hilbert space, H, and which may depend upon $u(x, t)$. (By L_t we mean the derivative with respect to the parameter t as it appears *explicitly* in the operator L; for example, if

$$L = -\frac{\partial^2}{\partial x^2} + u(x, t) \text{ then } L_t = u_t.)$$

The Hilbert space, H, is a space with an inner product, (ϕ, ψ), which is complete; we assume that L is self-adjoint, so that $(L\phi, \psi) = (\phi, L\psi)$ for all $\phi, \psi \in H$.

Now we introduce the *eigenvalue* (or *spectral*) *equation*, for $\psi \in H$,

$$L\psi = \lambda\psi \qquad \text{for } t \geqslant 0 \text{ and } -\infty < x < \infty, \tag{5.25}$$

where $\lambda = \lambda(t)$. Differentiating with respect to t, we see that

$$L_t\psi + L\psi_t = \lambda_t\psi + \lambda\psi_t$$

which becomes, upon the use of equation (5.24),

$$\begin{aligned}
\lambda_t\psi &= (L - \lambda)\psi_t + (ML - LM)\psi \\
&= (L - \lambda)\psi_t + M\lambda\psi - LM\psi \\
&= (L - \lambda)(\psi_t - M\psi).
\end{aligned} \tag{5.26}$$

The inner product of ψ with this equation gives

$$\begin{aligned}
(\psi, \psi)\lambda_t &= (\psi, (L - \lambda)(\psi_t - M\psi)) \\
&= ((L - \lambda)\psi, \psi_t - M\psi)
\end{aligned}$$

since $L - \lambda$ is self-adjoint, and so

$$(\psi, \psi)\lambda_t = (0, \psi_t - M\psi) = 0$$

or

$$\lambda_t = 0.$$

Thus each eigenvalue of the operator L is a constant.

From equation (5.26), with $\lambda_t = 0$, we obtain

$$(L - \lambda)(\psi_t - M\psi) = 0,$$

so that $\psi_t - M\psi$ is an eigenfunction of the operator L with eigenvalue λ. Hence

$$\psi_t - M\psi \propto \psi,$$

and we can always redefine M with the addition of the product of the identity operator and an appropriate function of t; this will not alter equation (5.24) (which is a representation of equation (5.23)). Thus we have the *time-evolution equation* for ψ,

$$\psi_t = M\psi, \qquad \text{for } t > 0. \tag{5.27}$$

In other words we have the following theorem:

If the evolution equation

$$u_t - N(u) = 0$$

can be expressed as the *Lax equation*

$$L_t + LM - ML = L_t + [L, M] = 0, \tag{5.28}$$

where $[L, M] = LM - ML$ is the *commutator* of the operators L and M, and if

$$L\psi = \lambda\psi,$$

then $\lambda_t = 0$ and ψ evolves according to equation (5.27).

Of course we recognise the choice

$$L = -\frac{\partial^2}{\partial x^2} + u,$$

and M is presumably associated with the operator appearing in the definition of R (see equations (4.1), (4.18), (4.23) and (4.25)); the connection with the KdV equation seems tenable, but clearly some further investigation is necessary in order to fill in the details.

5.2.2 The Lax KdV hierarchy

The problem we shall address here is the obvious one: how to choose the operators L and M, which satisfy the conditions laid down in the previous section. For our present purpose, it is convenient to restrict the discussion somewhat. We shall consider only that spectral equation which is given by the Schrödinger operator, i.e.

$$L = -\frac{\partial^2}{\partial x^2} + u, \tag{5.29}$$

so that $L\psi = \lambda\psi$ becomes the Sturm–Liouville equation. (Of course, the L given in (5.29) is self-adjoint.) With L chosen, the problem becomes one of determining M only.

Now M is a linear operator (in x) which, it turns out (see Q5.11), is to be anti- (or 'skew'-) symmetric so that $(M\phi, \psi) = (\phi, M\psi)$ for all $\phi, \psi \in H$; a natural choice is therefore to construct M from a suitable linear combination of *odd* derivatives in x. (This is quite easy to see: consider the inner product $(\phi, \psi) = \int_{-\infty}^{\infty} \phi\psi \, dx$, then

$$(M\phi, \psi) = \int_{-\infty}^{\infty} \frac{\partial^n \phi}{\partial x^n} \psi \, dx = - \int_{-\infty}^{\infty} \phi \frac{\partial^n \psi}{\partial x^n} \, dx = -(\phi, M\psi),$$

if $M = \partial^n/\partial x^n$ for n odd and $\phi, \phi_x, \ldots, \psi, \psi_x \cdots \to 0$ as $x \to \pm \infty$.) Further, M is required to have enough freedom in any unknown constants or functions to enable the operator $L_t + [L, M]$ to be chosen so that it is of degree zero, i.e. a multiplicative operator. The simplest choice is obviously

$$M = c \frac{\partial}{\partial x}, \tag{5.30}$$

where we shall take c to be a constant, and then

$$[L, M] = c\left(-\frac{\partial^2}{\partial x^2} + u\right)\frac{\partial}{\partial x} - c\frac{\partial}{\partial x}\left(-\frac{\partial^2}{\partial x^2} + u\right) = -cu_x,$$

which is automatically a multiplicative operator! Thus

$$L_t + [L, M] = u_t - cu_x,$$

and so the one-dimensional wave equation

$$u_t - cu_x = 0 \tag{5.31}$$

has an associated spectral problem with eigenvalues which are constants of the motion. This is not very useful since we can solve the initial-value problem for equation (5.31) using direct methods. (Note that if we set $c = c(x, t)$ in the above, then $L_t + [L, M]$ is of degree zero only if $c = c(t)$; this is left as an exercise.)

The next level of complexity is to choose M so that it involves, at most, a third-order differential operator,

$$M = -\alpha \frac{\partial^3}{\partial x^3} + U \frac{\partial}{\partial x} + \frac{\partial}{\partial x} U + A, \tag{5.32}$$

where α is a constant, $U = U(x, t)$ and $A = A(x, t)$. (We could just as well write

$$M = -\alpha \frac{\partial^3}{\partial x^3} + B(x, t)\frac{\partial}{\partial x} + C(x, t),$$

but the pattern evident in (5.32) is useful in the later discussion.) It then follows, after an elementary calculation, that

$$[L, M] = \alpha u_{xxx} - U_{xxx} - A_{xx} - 2u_x U + (3\alpha u_{xx} - 4U_{xx} - 2A_x)\frac{\partial}{\partial x}$$

$$+ (3\alpha u_x - 4U_x)\frac{\partial^2}{\partial x^2},$$

which is a multiplicative operator if

$$U = \tfrac{3}{4}\alpha u \qquad \text{and} \qquad A = A(t).$$

Thus the Lax equation, (5.28), becomes

$$u_t + \tfrac{1}{4}\alpha u_{xxx} - \tfrac{3}{2}\alpha u u_x = 0,$$

which is our conventional KdV equation if $\alpha = 4$. The operator M, defined by equation (5.32), is therefore

$$M = -4\frac{\partial^3}{\partial x^3} + 3u\frac{\partial}{\partial x} + 3\frac{\partial}{\partial x}u + A(t)$$

and so the time-evolution equation for ψ is

$$\psi_t = -4\psi_{xxx} + 3u\psi_x + 3(u\psi)_x + A\psi.$$

This equation can be recast, on using the Sturm–Liouville equation

$$\psi_{xx} + (\lambda - u)\psi = 0,$$

as

$$\psi_t = 4(\lambda\psi - u\psi)_x + 3u\psi_x + 3(u\psi)_x + A\psi$$
$$= 2(u + 2\lambda)\psi_x - u_x\psi + A\psi$$

which, for $A = 0$, is the time-evolution equation, (4.23), for the discrete eigenfunctions; with $A = 4ik^3$ we have the corresponding equation for the continuous eigenfunction (see equation (4.25) with equation (4.28)).

The KdV equation has appeared as the second example within the framework of the Lax formulation, on having chosen L to be the Schrödinger operator. It is however – as we noted earlier – the first non-trivial case. The procedure adopted above can now be extended to higher-order nonlinear evolution equations. A little trial and error (or experience with commutators) shows that an appropriate choice for M is

$$M = -\alpha\frac{\partial^{2n+1}}{\partial x^{2n+1}} + \sum_{m=1}^{n}\left(U_m\frac{\partial^{2m-1}}{\partial x^{2m-1}} + \frac{\partial^{2m-1}}{\partial x^{2m-1}}U_m\right) + A \qquad (5.33)$$

where α is a constant, $U_m = U_m(x, t)$ and $A = A(t)$. (We note that $A(t)$ plays no rôle in the resulting evolution equation, and α is introduced purely for convenience.) The requirement that $[L, M]$ is to be a multiplicative operator imposes n conditions on the n unknown functions U_m. The case

$n = 1$ generates the KdV equation (see equation (5.32)), and for $n = 2$ it can be shown that the evolution equation is

$$u_t + 30u^2 u_x - 20u_x u_{xx} - 10uu_{xxx} + u_{xxxxx} = 0, \tag{5.34}$$

a fifth-order KdV equation. We have now presented the first three evolution equations in the *KdV hierarchy*, every one of which can be solved by the inverse scattering transform method described in Chap. 4.

We next introduce an important connection with the Hamiltonian formulation that was mentioned in §5.1.3. The equations in the KdV hierarchy can also be obtained from

$$k\frac{\partial u}{\partial t} = \frac{\partial}{\partial x}\left(\frac{\delta T_n}{\delta u}\right), \qquad n = 3, 4, \ldots,$$

where $\delta/\delta u$ is the variational derivative, $k(\neq 0)$ is an arbitrary real constant and the T_ns are given in §5.1.1. Thus T_3 generates the KdV equation (as we have already seen), T_4 gives equation (5.34) and so on (see Q5.15). The constant k can be removed by replacing t by kt, an elementary rescaling of t.

Clearly, there is an infinity of evolution equations in this hierarchy. Furthermore, the number of equations generated by the Lax formalism could be extended in two obvious ways. First, we could investigate other spectral equations by choosing alternative forms for L. Second, there is no requirement that L and M should be restricted to the class of scalar operators; L and M could be *matrix operators*. This final point turns out to be a particularly powerful idea; we shall discuss the matrix approach in the next chapter.

5.3 Hirota's method: the bilinear form

In the previous chapter we discussed the N-soliton solution of the KdV equation (see example (iii), §4.5). We found, in particular, that this solution could be expressed in the form

$$u = -2\frac{\partial^2}{\partial x^2}\log f, \tag{5.35}$$

where $f(x, t)$ is the determinant of the appropriate matrix. It is thence possible to discuss the solution, u, of the KdV equation by transforming the dependent variable u to f, although this might produce an even more complicated problem. Hirota, starting in 1971, followed this approach in a sequence of papers which dealt not only with the KdV equation but also with many other nonlinear evolution equations. He found that, under a suitable transformation (analogous to (5.35)), an evolution equation could eventually be reduced to a *bilinear form*, a version of the original equation which requires the introduction of a novel differential operator. We shall describe the general method in the context of the KdV equation, and then derive its two-soliton solution.

5.3.1 The bilinear operator

First we construct the equation for $f(x, t)$, using the transformation (5.35), where $u(x, t)$ satisfies the KdV equation

$$u_t - 6uu_x + u_{xxx} = 0,$$

with $f_x, f_{xx}, \cdots \to 0$ as $x \to +\infty$ or $x \to -\infty$. (This decay condition is equivalent to that imposed at infinity for the soliton solutions, although an extension of the bilinear method allows periodic solutions also to be obtained; see Matsuno (1984, §2.2.3).) Substitution for u into the KdV equation yields, after a little manipulation and one integration in x, the equation for f

$$ff_{xt} - f_x f_t + ff_{xxxx} - 4f_x f_{xxx} + 3f_{xx}^2 = 0, \tag{5.36}$$

where the condition at $+\infty$ or $-\infty$ has been used. (It is usually simpler to first set $u = w_x$, integrate once with $w_t, w_x, \cdots \to 0$ as $x \to \pm\infty$, and then introduce $w = -2f_x/f$ which is essentially the Hopf–Cole transformation: see Q5.18.) The problem now is how to solve equation (5.36), an equation which appears a little more difficult than the original KdV equation! However, a related question – how best to express this new equation which is solely quadratic in f – allows us to make some headway.

Hirota introduced the *bilinear operator*, $D_t^m D_x^n(a \cdot b)$, defined as

$$D_t^m D_x^n(a \cdot b) = \left(\frac{\partial}{\partial t} - \frac{\partial}{\partial t'}\right)^m \left(\frac{\partial}{\partial x} - \frac{\partial}{\partial x'}\right)^n a(x, t)b(x', t')\bigg|_{\substack{x'=x \\ t'=t}} \tag{5.37}$$

for non-negative integers m and n. To see how this works, let us consider the case of $m = n = 1$ for which

$$\left(\frac{\partial}{\partial t} - \frac{\partial}{\partial t'}\right)\left(\frac{\partial}{\partial x} - \frac{\partial}{\partial x'}\right) a(x, t)b(x', t') = \left(\frac{\partial}{\partial t} - \frac{\partial}{\partial t'}\right)(a_x b - ab_{x'})$$

$$= a_{xt}b - a_t b_{x'} - a_x b_{t'} + ab_{x't'}.$$

When this is evaluated on $x' = x, t' = t$ we obtain

$$D_t D_x(a \cdot b) = a_{xt}b + ab_{xt} - a_t b_x - a_x b_t,$$

and if we choose $a = b$ for all x, t, then

$$D_t D_x(a \cdot a) = 2(aa_{xt} - a_x a_t), \tag{5.38}$$

which is to be compared with the two time-derivative terms appearing in equation (5.36).

(The bilinear operator has a simple analogue in elementary differentiation theory: if we consider

$$\left(\frac{\partial}{\partial x} + \frac{\partial}{\partial x'}\right) a(x, t) b(x', t') = a_x b + ab_{x'},$$

then on $x' = x, t' = t$ we obtain

$$\left(\frac{\partial}{\partial x} + \frac{\partial}{\partial x'}\right) a(x, t) b(x', t') \bigg|_{\substack{x'=x \\ t'=t}} = \frac{\partial}{\partial x}(ab).$$

Generalising this result, we find that

$$\left(\frac{\partial}{\partial t} + \frac{\partial}{\partial t'}\right)^m \left(\frac{\partial}{\partial x} + \frac{\partial}{\partial x'}\right)^n a(x, t) b(x', t') \bigg|_{\substack{x'=x \\ t'=t}} = \frac{\partial^{m+n}}{\partial t^m \partial x^n}(ab),$$

the standard derivative of a product.)

As an example with a higher derivative, let us find $D_x^4(a \cdot b)$, i.e. take $m = 0$ and $n = 4$. In this case we have

$$\left(\frac{\partial}{\partial x} - \frac{\partial}{\partial x'}\right)^4 a(x, t) b(x', t')$$

$$= a_{xxxx}b - 4a_{xxx}b_{x'} + 6a_{xx}b_{x'x'} - 4a_x b_{x'x'x'} + ab_{x'x'x'x'}.$$

If we now evaluate on $x' = x, t' = t$, and again make the special choice $a = b$ for all x, t, we obtain

$$D_x^4(a \cdot a) = 2(aa_{xxxx} - 4a_x a_{xxx} + 3a_{xx}^2). \tag{5.39}$$

We can see immediately that, if we compare equations (5.38) and (5.39) with equation (5.36), then the equation for f can be expressed as

$$D_x(D_t + D_x^3)(f \cdot f) = 0, \tag{5.40}$$

the bilinear form of the KdV equation. (Note here that if we *interpret* D_t and D_x as $\partial/\partial t$ and $\partial/\partial x$, respectively, then $D_t + D_x^3$ becomes the linearised operator in the KdV equation, obtained by letting $u \to 0$.) Now this compact equation looks deceptively simple, but it must be remembered that D represents a rather unconventional differential operator. We shall discuss the method of solution of this equation in the next section.

Before we leave these introductory ideas, let us briefly indicate two generalisations. First, provided solutions exist to some bilinear equations, we could generate evolution equations by *starting* from appropriate bilinear forms. For example, looking at equation (5.40), we might suggest the equation

$$\mathbf{D}_x(\mathbf{D}_t + \mathbf{D}_x^5)(f \cdot f) = 0 \qquad (5.41)$$

(together with the transformation (5.35)) as the next equation in a KdV hierarchy. (The connection with KdV equations is clear: equation (5.41) must give rise to a term u_t and also a term u_{xxxxx}, together with some mix of nonlinear terms.) In fact equations (5.41) and (5.35) yield

$$u_t + 45u^2 u_x - 15u_x u_{xx} - 15uu_{xxx} + u_{xxxxx} = 0, \qquad (5.42)$$

the *Sawada–Kotera equation* (Sawada & Kotera, 1974), which is to be compared with equation (5.34). Now we have a surprise: these two fifth-order KdV equations, (5.34) and (5.42), are *not* equivalent, i.e. there is no scaling transformation which takes one into the other (see Q5.19). Thus we can construct *two* KdV hierarchies, both of which contain the classical KdV equation as one of their members, and it turns out that each member in both hierarchies can be solved by the inverse scattering transform.

The second generalisation is the more difficult, and yet it is the more natural one to explore. That is, given an evolution equation, how can it be expressed in bilinear form (assuming such a form exists)? We shall mention three examples (and leave a few others, and the details, as exercises at the end of the chapter). An equation which describes one-dimensional weakly nonlinear dispersive water waves, for waves propagating in *both* directions, is

$$u_{tt} - u_{xx} + 3(u^2)_{xx} - u_{xxxx} = 0, \qquad (5.43)$$

the *Boussinesq equation* (Hirota, 1973b). (A suitable approximation enables the KdV equation to be derived from this equation, if the wave motion is restricted to be unidirectional.) Equation (5.43) has the bilinear form (see Q5.21)

$$(\mathbf{D}_t^2 - \mathbf{D}_x^2 - \mathbf{D}_x^4)(f \cdot f) = 0, \qquad (5.44)$$

where u and f are related by the usual transformation

$$u(x, t) = -2\frac{\partial^2}{\partial x^2}\log f. \qquad (5.45)$$

Again, we note the correspondence between equation (5.44) and the linearised version of equation (5.43).

It is clear that the transformation (5.45) might not be the appropriate one to choose for certain evolution equations. Consider $u(x, t)$ which satisfies the *Benjamin–Ono equation*

$$u_t + 4uu_x + \mathcal{H}(u_{xx}) = 0, \qquad (5.46)$$

where $\mathcal{H}$ is the Hilbert transform

$$\mathcal{H}(w) = \frac{1}{\pi} \int_{-\infty}^{\infty} \frac{w(y,t)}{y-x} \, dy,$$

(see Q1.14 and Q1.15). If we introduce the transformation

$$u(x,t) = \frac{i}{2} \frac{\partial}{\partial x} \log(f^*/f), \tag{5.47}$$

where $f(x,t)$ is a complex-valued function, and the asterisk denotes the complex conjugate, then the bilinear form of equation (5.46) turns out to be

$$(\mathrm{i}D_t - D_x^2)(f^* \cdot f) = 0; \tag{5.48}$$

see Q5.22.

Finally, let us return to the fifth-order KdV equation, (5.34), which is a member of the 'Lax hierarchy'. This equation does have a bilinearisation, but in this case a *pair* of bilinear equations is necessary. If u and f are related by the standard transformation, (5.45), and if we introduce an auxiliary independent variable, τ say, then the bilinear form of

$$u_t + 30u^2 u_x - 20u_x u_{xx} - 10uu_{xxx} + u_{xxxxx} = 0$$

is

$$\left.\begin{array}{r} \{D_x(D_t + D_x^5) - \tfrac{5}{6}D_x^3(D_\tau + D_x^3)\}(f \cdot f) = 0 \\ D_x(D_\tau + D_x^3)(f \cdot f) = 0; \end{array}\right\} \tag{5.49}$$

see Q5.27 for the details.

5.3.2 The solution of the bilinear equation

We now turn to the problem of how to solve bilinear equations. In particular, we shall discuss the KdV bilinear form,

$$D_x(D_t + D_x^3)(f \cdot f) = 0, \tag{5.50}$$

although the method we shall describe is generally applicable. The method requires the use of some properties of the bilinear operator; these will be introduced below, but the relevant derivations are left as exercises at the end of the chapter.

Our starting point is the solitary-wave solution of the KdV equation

$$u(x,t) = -2\,\mathrm{sech}^2(x - 4t)$$

(see equation (4.34)), which can be written as

$$u(x,t) = 4\frac{\partial}{\partial x}\left(\frac{e^{8t-2x}}{1 + e^{8t-2x}}\right) = -2\frac{\partial^2}{\partial x^2}\log(1 + e^{8t-2x}).$$

Thus the solitary wave, of amplitude -2, is represented by the choice

$$f(x, t) = 1 + e^{8t - 2x}. \tag{5.51}$$

First we confirm that f, as given by (5.51), is a solution of the bilinear equation, (5.50). We note that

$$D_t D_x(a \cdot 1) = \frac{\partial^2 a}{\partial t \partial x} = D_t D_x(1 \cdot a),$$

and

$$D_t^m D_x^n \{\exp(\theta_1) \cdot \exp(\theta_2)\} = (\omega_2 - \omega_1)^m (k_1 - k_2)^n \exp(\theta_1 + \theta_2)$$

where $\theta_i = k_i x - \omega_i t + \alpha_i$ (see Q5.20). Thus, with f given by (5.51), we have

$$B(f \cdot f) = B(1 \cdot 1) + B(1 \cdot e^{8t - 2x}) + B(e^{8t - 2x} \cdot 1) + B(e^{8t - 2x} \cdot e^{8t - 2x})$$

where B is any bilinear operator, and so we obtain

$$D_x(D_t + D_x^3)(f \cdot f) = -32e^{8t - 2x} + 2(-2)^4 e^{8t - 2x} = 0;$$

f, from (5.51), is indeed a solution of equation (5.50).

The solitary-wave solution, expressed by equation (5.51), is now generalised to accommodate the N-soliton solution. This is most easily accomplished by the introduction of an arbitrary parameter ε, together with the assumption that f may be expanded in integral powers of ε. (It will turn out that the series we generate will terminate after a finite number of terms, and so ε may be assigned arbitrarily.) Let us write

$$f = 1 + \sum_{n=1}^{\infty} \varepsilon^n f_n(x, t), \tag{5.52}$$

and substitute directly into the bilinear equation (5.50). If we collect like powers of ε, we obtain

$$B(1 \cdot 1) + \varepsilon B(f_1 \cdot 1 + 1 \cdot f_1) + \varepsilon^2 B(f_2 \cdot 1 + f_1 \cdot f_1 + 1 \cdot f_2) + \cdots$$

$$+ \varepsilon^r B\left(\sum_{m=0}^{r} f_{r-m} \cdot f_m\right) + \cdots = 0 \tag{5.53}$$

where $f_0 = 1$ and the bilinear operator here is

$$B = D_x(D_t + D_x^3).$$

Now $B(1 \cdot 1) = 0$, and each coefficient of $\varepsilon^r (r = 1, 2, \ldots)$ must be zero so that

$$B(f_1 \cdot 1 + 1 \cdot f_1) = 0,$$

and so on. This equation for f_1 reduces to

$$\left(\frac{\partial}{\partial t} + \frac{\partial^3}{\partial x^3}\right) f_1 = 0,$$

since $B(a \cdot b + c \cdot d) = B(a \cdot b) + B(c \cdot d)$, provided $f_{1t}, f_{1x}, \cdots \to 0$ as $x \to +\infty$

or $-\infty$. We shall write this equation as

$$\mathrm{D}f_1 = 0 \qquad \text{with } \mathrm{D} = \frac{\partial}{\partial t} + \frac{\partial^3}{\partial x^3}. \tag{5.54}$$

The next two equations in this sequence can then be expressed as

$$2\mathrm{D}f_2 = -\mathrm{B}(f_1 \cdot f_1) \qquad \text{and} \qquad 2\mathrm{D}f_3 = -\mathrm{B}(f_1 \cdot f_2 + f_2 \cdot f_1). \tag{5.55}$$

It is easily checked that, if $f_1 = \exp(\theta_1)$ where $\theta_i = k_i x - k_i^3 t + \alpha_i$ with k_i and α_i arbitrary constants, then

$$\mathrm{D}f_1 = 0 \text{ and, further, that } f_2 \text{ satisfies } \mathrm{D}f_2 = 0.$$

Hence we may choose $f_n = 0$, $n = 2, 3, \ldots$, for all x, t, and the resulting form of $f(x, t)$ recovers the solitary wave. Furthermore, since the series (5.52) terminates after $n = 1$, we may set $\varepsilon = 1$ without loss of generality. (We see that f_1 already incorporates an arbitrary constant phase shift by virtue of α_1.)

The equation for f_1, (5.54), is linear and so we may add any number of exponential terms: let us suppose that

$$f_1 = \exp(\theta_1) + \exp(\theta_2) \tag{5.56}$$

where θ_i is defined above. As before, $\mathrm{D}f_1 = 0$, but now we obtain from the first of equations (5.55)

$$\begin{aligned}
2\mathrm{D}f_2 &= -\mathrm{B}(f_1 \cdot f_1) \\
&= -\mathrm{B}\{\exp(\theta_1) \cdot \exp(\theta_1)\} - \mathrm{B}\{\exp(\theta_1) \cdot \exp(\theta_2)\} \\
&\quad - \mathrm{B}\{\exp(\theta_2) \cdot \exp(\theta_1)\} - \mathrm{B}\{\exp(\theta_2) \cdot \exp(\theta_2)\}
\end{aligned}$$

and the only non-zero terms are those involving both θ_1 and θ_2; thus we find that

$$2\mathrm{D}f_2 = -2\{(k_1 - k_2)(\omega_2 - \omega_1) + (k_1 - k_2)^4\} \exp(\theta_1 + \theta_2), \tag{5.57}$$

where $\omega_i = k_i^3$. This equation (5.57), for $f_2(x, t)$, has a particular integral of the form

$$f_2 = A_2 \exp(\theta_1 + \theta_2), \tag{5.58}$$

and direct substitution yields

$$A_2 = \left(\frac{k_1 - k_2}{k_1 + k_2}\right)^2.$$

The second of equations (5.55) now becomes

$$\begin{aligned}
2\mathrm{D}f_3 = &-A_2 \mathrm{B}\{\exp(\theta_1) \cdot \exp(\theta_1 + \theta_2)\} - A_2 \mathrm{B}\{\exp(\theta_1 + \theta_2) \cdot \exp(\theta_1)\} \\
&- A_2 \mathrm{B}\{\exp(\theta_2) \cdot \exp(\theta_1 + \theta_2)\} - A_2 \mathrm{B}\{\exp(\theta_1 + \theta_2) \cdot \exp(\theta_2)\}
\end{aligned}$$

which can be written as

$$2Df_3 = -2A_2\{(-k_2)(k_2^3) + (-k_2)^4\}\exp(2\theta_1 + \theta_2)$$
$$- 2A_2\{(-k_1)(k_1^3) + (-k_1)^4\}\exp(2\theta_2 + \theta_1)$$
$$= 0.$$

So this time, for f_1 given by equation (5.56) and f_2 defined by (5.58), we may choose $f_n = 0$, $n = 3, 4, \ldots$, for all x and t, and then with $\varepsilon = 1$ we have another exact solution

$$f = 1 + \exp(\theta_1) + \exp(\theta_2) + \left(\frac{k_1 - k_2}{k_1 + k_2}\right)^2 \exp(\theta_1 + \theta_2); \qquad (5.59)$$

this generates the general two-soliton solution of the KdV equation (cf. Q4.3).

The method of solution described above can be extended by writing

$$f_1 = \sum_{i=1}^{N} \exp(\theta_i). \qquad (5.60)$$

It can then be shown that the series (5.52) terminates after the term f_N; the series generates the N-soliton solution. The construction of this solution, although fairly straightforward, is tedious and will not be pursued here. However, the form of f for the three-soliton solution is set as an exercise (see Q5.29). The solution expressed in this way, like the two-soliton solution written explicitly in equation (5.59), represents a *nonlinear superposition principle* for the construction of the N-soliton solution of the KdV equation.

5.4 Bäcklund transformations

Bäcklund transformations were developed in the 1880s for use in the related theories of differential geometry and differential equations. They arose as a generalisation of *contact transformations*, i.e. transformations that take surfaces with a common tangent at a point in one space into surfaces, in another space, which also have a common tangent at the corresponding point. One of the earliest Bäcklund transformations was for the *sine–Gordon equation*, sometimes called the *SG equation*,

$$u_{xt} = \sin u,$$

an equation originally arising in differential geometry to describe surfaces with a constant negative Gaussian curvature (Enneper, 1870). The sine–Gordon equation, so named as a pun on Klein–Gordon, can also be solved by the inverse scattering transform (see Chap. 6). Indeed, there would seem to be a close relationship between the inverse scattering transform (IST) and a Bäcklund transformation (BT): every evolution equation solvable by IST has a corresponding BT.

In this section we shall describe the general character of the Bäcklund transformation, with examples, and then apply the idea specifically to the KdV equation.

5.4.1 Introductory ideas

Suppose that we have two uncoupled partial differential equations, in two independent variables x and t, for the two functions u and v; the two equations are expressed as

$$P(u) = 0 \qquad \text{and} \qquad Q(v) = 0, \tag{5.61}$$

where P and Q are two operators, which are in general nonlinear. Let $R_i = 0$ be a pair of relations,

$$R_i(u, v, u_x, v_x, u_t, v_t, \ldots; x, t) = 0, \qquad i = 1, 2, \tag{5.62}$$

between the two functions u and v. Then $R_i = 0$ is a *Bäcklund transformation* if it is integrable for v when $P(u) = 0$ and if the resulting v is a solution of $Q(v) = 0$, and vice versa. If $P = Q$, so that u and v satisfy the *same* equation, then $R_i = 0$ is called an *auto-Bäcklund transformation*. Of course, this approach to the solution of the equations $P(u) = 0$ and $Q(v) = 0$ is normally only useful if the relations $R_i = 0$ are, in some sense, simpler than the original equations (5.61).

One of the simplest auto-Bäcklund transformations is the pair (written with y rather than t)

$$u_x = v_y, \qquad u_y = -v_x,$$

the Cauchy–Riemann relations for Laplace's equation

$$u_{xx} + u_{yy} = 0; \qquad v_{xx} + v_{yy} = 0.$$

Thus, if $v(x, y) = xy$ (a simple solution of Laplace's equation), then $u(x, y)$ can be determined from

$$u_x = x \qquad \text{and} \qquad u_y = -y,$$

and so $u(x, y) = \frac{1}{2}(x^2 - y^2)$ is another solution of Laplace's equation.

Another fairly simple application arises in connection with *Liouville's equation*, which we shall write as

$$u_{xt} = e^u. \tag{5.63}$$

First we introduce an auxiliary dependent variable, v, which satisfies

$$v_{xt} = 0. \tag{5.64}$$

Now let us consider the pair of first-order equations

$$u_x + v_x = \sqrt{2}e^{(u-v)/2} \quad \text{and} \quad u_t - v_t = \sqrt{2}e^{(u+v)/2}, \quad (5.65)$$

and cross-differentiate to obtain

$$u_{xt} + v_{xt} = \frac{1}{\sqrt{2}}(u_t - v_t)e^{(u-v)/2} = e^u$$

and
$$\left. \begin{array}{c} \\ \end{array} \right\} \quad (5.66)$$

$$u_{tx} - v_{tx} = \frac{1}{\sqrt{2}}(u_x + v_x)e^{(u+v)/2} = e^u.$$

It is immediately clear that the two equations (5.66) imply equations (5.63) and (5.64); thus the pair of equations (5.65) constitute a Bäcklund transformation for Liouville's equation and the equation $v_{xt} = 0$. Now this latter equation is easily solved and so, from the Bäcklund transformation (5.65), we can generate the general solution of Liouville's equation; this is left as an exercise (see Q5.31).

Finally, in this introduction, let us return to the sine–Gordon equation written in characteristic form

$$u_{xt} = \sin u, \quad (5.67)$$

which might reasonably be regarded as the forerunner of equations with a Bäcklund transformation (although the form we shall discuss here is probably due to Bianchi). Consider the pair of equations

$$\tfrac{1}{2}(u+v)_x = a\sin\left(\frac{u-v}{2}\right); \quad \tfrac{1}{2}(u-v)_t = \frac{1}{a}\sin\left(\frac{u+v}{2}\right), \quad (5.68)$$

where $a \ (\neq 0)$ is an arbitrary real constant. Again, we form the cross-derivatives

$$\tfrac{1}{2}(u+v)_{xt} = \frac{a}{2}(u-v)_t\cos\left(\frac{u-v}{2}\right) = \sin\left(\frac{u+v}{2}\right)\cos\left(\frac{u-v}{2}\right)$$

and
$$\left. \begin{array}{c} \\ \end{array} \right\} \quad (5.69)$$

$$\tfrac{1}{2}(u-v)_{tx} = \frac{1}{2a}(u+v)_x\cos\left(\frac{u+v}{2}\right) = \sin\left(\frac{u-v}{2}\right)\cos\left(\frac{u+v}{2}\right);$$

thus we see that

$$u_{xt} = \sin u \quad \text{and} \quad v_{xt} = \sin v \quad (5.70)$$

(by adding and subtracting, respectively, the equations (5.69)). Since both u and v satisfy the sine–Gordon equation, equations (5.68) are an auto-Bäcklund transformation for equation (5.67).

To see how we can use this transformation, let us adopt a manoeuvre which is particularly suited to solving evolution equations. It is obvious that the sine–Gordon equation, (5.67), has the zero solution, $u(x, t) = 0$, for all x and t. We can now use this trivial solution to generate a non-trivial one. Let us choose $v = 0$, then the Bäcklund transformation, (5.68), becomes

$$u_x = 2a \sin(\tfrac{1}{2}u) \qquad \text{and} \qquad u_t = (2/a) \sin(\tfrac{1}{2}u),$$

and these two equations may be integrated to give

$$2ax = \int^u \frac{du}{\sin(\tfrac{1}{2}u)} = 2 \log|\tan \tfrac{1}{4}u| + f(t)$$

and

$$2t/a = 2 \log|\tan \tfrac{1}{4}u| + g(x),$$

respectively, where f and g are arbitrary functions. Thus, for consistency, we must have

$$\tan(\tfrac{1}{4}u) = C \exp(ax + t/a)$$

or

$$u(x, t) = 4 \arctan\{C \exp(ax + t/a)\}, \tag{5.71}$$

where C is an arbitrary constant. Since both u and v are solutions of the sine–Gordon equation, we have obtained a new solution given by (5.71); in fact equation (5.71) describes the solitary-wave solution of the sine–Gordon equation (see Chap. 6 and Q2.19).

5.4.2 Bäcklund transformation for the KdV equation

We have already introduced the Miura transformation (4.8),

$$u = v^2 + v_x \tag{5.72}$$

and shown that, if v is a solution of the modified KdV equation

$$v_t - 6v^2 v_x + v_{xxx} = 0, \tag{5.73}$$

then u is a solution of the KdV equation

$$u_t - 6uu_x + u_{xxx} = 0. \tag{5.74}$$

Since we can eliminate higher derivatives from equation (5.73) by use of equation (5.72), we may regard equations (5.72) and (5.73) as a Bäcklund transformation for the KdV equation. However, a more convenient transformation was developed by Wahlquist & Estabrook (1973), which we shall now describe.

First, we recall that the KdV equation is Galilean invariant and so we could work with $u - \lambda$ rather than u; this we choose to do. Thus we write

$$u = \lambda + v^2 + v_x, \tag{5.75}$$

where λ is a real parameter, and then the modified KdV equation becomes

$$v_t - 6(v^2 + \lambda)v_x + v_{xxx} = 0 \qquad (5.76)$$

so that equations (5.76) and (5.75) imply the original KdV equation, (5.74), for u. Further, we see that if v is a solution of equation (5.76) then so is $-v$; this suggests that we introduce two functions, u_1 and u_2, defined by

$$u_1 = \lambda + v^2 + v_x; \qquad u_2 = \lambda + v^2 - v_x,$$

for given v and λ. These two equations imply that

$$u_1 - u_2 = 2v_x \qquad \text{and} \qquad u_1 + u_2 = 2(\lambda + v^2). \qquad (5.77)$$

At this stage it is convenient to introduce the additional transformation

$$u_i = \frac{\partial w_i}{\partial x} \qquad (i = 1, 2), \qquad (5.78)$$

so that equations (5.77) become

$$w_1 - w_2 = 2v \qquad (5.79)$$

and

$$(w_1 + w_2)_x = 2\lambda + \tfrac{1}{2}(w_1 - w_2)^2, \qquad (5.80)$$

respectively. (Note that equation (5.79) is obtained by one integration in x, but the resulting arbitrary function in t can always be absorbed into the definition of w_i without altering u_i.) Equation (5.80) constitutes part of the Bäcklund transformation for w_1 and w_2 which, in turn, generate solutions of the KdV equation via equation (5.78). Finally, equation (5.76) can be written as

$$(w_1 - w_2)_t - 3(w_{1x}^2 - w_{2x}^2) + (w_1 - w_2)_{xxx} = 0, \qquad (5.81)$$

by using equations (5.77)–(5.79). This equations (5.80) and (5.81) (with (5.78)) constitute the auto-Bäcklund transformation for the KdV equation, where equation (5.80) describes the 'x' part of the transformation and equation (5.81) the 't' part.

To give an example of the use of equations (5.80) and (5.81) we shall employ the same procedure as for the sine–Gordon equation (described in §5.4.1). Thus we start from the zero solution: let $w_2(x, t) = 0$ for all x and t; then equation (5.80) becomes

$$w_{1x} = 2\lambda + \tfrac{1}{2}w_1^2, \qquad (5.82)$$

which can be integrated directly to yield

$$w_1(x, t) = -2\kappa \tanh\{\kappa x + f(t)\} \qquad (5.83)$$

where $\lambda = -\kappa^2 \ (< 0)$ and f is an arbitrary function. The dependence on

hand side of the resulting equation

$$0 = 4(\lambda_2 - \lambda_1) + \tfrac{1}{2}\{(w_{12} - w_1)^2 - (w_{21} - w_2)^2 - (w_1 - w_0)^2 + (w_2 - w_0)^2\}.$$

Again we use equation (5.90), and solve for w_{12} $(= w_{21})$, to give

$$w_{12} = w_0 - \frac{4(\lambda_1 - \lambda_2)}{w_1 - w_2}, \tag{5.91}$$

which describes w_{12} in simple *algebraic* terms. Thus it is now possible to construct solutions of the KdV equation in a remarkably straightforward manner, although the process is not entirely labour-free. We observe that equation (5.91) constitutes a *nonlinear superposition principle* for the generation of solutions of the KdV equation (cf. equation (5.59)).

As an example of the use of equation (5.91), let us take $w_0 = 0$, $w_1 = -2\tanh(x - 4t)$ and $w_2 = -4\coth(2x - 32t)$, so that λ_1 and λ_2 are -1 and -4, respectively; thus we obtain

$$w_{12}(x, t) = -6/\{2\coth(2x - 32t) - \tanh(x - 4t)\}.$$

The corresponding solution of the KdV equation is obtained from $u_{12} = \partial w_{12}/\partial x$, so that

$$
\begin{aligned}
u_{12}(x, t) &= -6\frac{4\operatorname{cosech}^2(2x - 32t) + \operatorname{sech}^2(x - 4t)}{\{2\coth(2x - 32t) - \tanh(x - 4t)\}^2} \\
&= -6\frac{4\cosh^2(x - 4t) + \sinh^2(2x - 32t)}{\{2\cosh(2x - 32t)\cosh(x - 4t) - \sinh(x - 4t)\sinh(2x - 32t)\}^2} \\
&= -12\frac{3 + 4\cosh(2x - 8t) + \cosh(4x - 64t)}{\{3\cosh(x - 28t) + \cosh(3x - 36t)\}^2},
\end{aligned}
$$

which is the two-soliton solution (see equation (4.36)). It is interesting to note that this non-singular solution of the KdV equation requires a combination of both a non-singular *and* a singular solution in equation (5.91).

5.4.4 Bäcklund transformations and the bilinear form

The bilinear form of the KdV equation, for $f(x, t)$, is

$$D_x(D_t + D_x^3)(f \cdot f) = 0, \tag{5.92}$$

where

$$u(x, t) = -2\frac{\partial^2}{\partial x^2}\log f$$

(see §5.3.1). We shall now use the KdV equation as an example to demonstrate how the Bäcklund transformation can be recovered directly from the bilinear form. In keeping with the notion of a Bäcklund transformation, let us introduce two functions, f_1 and f_2, both of which are solutions of equation (5.92), so that

$$D_x(D_t + D_x^3)(f_1 \cdot f_1) = 0$$

and

$$D_x(D_t + D_x^3)(f_2 \cdot f_2) = 0.$$

Further, let us construct the equation

$$f_1^2 D_x(D_t + D_x^3)(f_2 \cdot f_2) - f_2^2 D_x(D_t + D_x^3)(f_1 \cdot f_1) = 0, \tag{5.93}$$

from which it is clear that, if f_1 satisfies equation (5.92) then so does f_2, and vice versa. Hence equation (5.93) must constitute an auto-Bäcklund transformation for the KdV equation. In order to write this transformation in the more conventional form (with an 'x' part and a 't' part) we must introduce a parameter λ. To this end we add

$$6\lambda D_x(A \cdot B) - 6\lambda D_x(A \cdot B) \tag{5.94}$$

to equation (5.93), where we choose

$$A = D_x(f_1 \cdot f_2) \qquad \text{and} \qquad B = f_1 f_2,$$

so that the expression (5.94) can be written as

$$- 6\lambda D_x[\{D_x(f_2 \cdot f_1)\} \cdot (f_1 f_2)] + 6\lambda D_x[(f_1 f_2) \cdot \{D_x(f_1 \cdot f_2)\}].$$

Equation (5.93) now becomes

$$f_1^2 D_x(D_t + D_x^3)(f_2 \cdot f_2) - 6\lambda D_x[\{D_x(f_2 \cdot f_1)\} \cdot (f_1 f_2)]$$
$$- f_2^2 D_x(D_t + D_x^3)(f_1 \cdot f_1) + 6\lambda D_x[(f_1 f_2) \cdot \{D_x(f_1 \cdot f_2)\}] = 0,$$

and on using the identities given in Q5.36 this becomes, after some manipulation,

$$2D_x[\{(D_t - 3\lambda D_x + D_x^3)(f_2 \cdot f_1)\} \cdot (f_1 f_2)]$$
$$+ 6D_x[\{(D_x^2 + \lambda)(f_2 \cdot f_1)\} \cdot \{D_x(f_1 \cdot f_2)\}] = 0. \tag{5.95}$$

(The details in the derivation of this latter equation are left as an exercise.)

It is evident that equation (5.95) is satisfied if

$$(D_t - 3\lambda D_x + D_x^3)(f_2 \cdot f_1) = 0 \tag{5.96}$$

and

$$(D_x^2 + \lambda)(f_2 \cdot f_1) = 0; \tag{5.97}$$

it is this pair of equations which corresponds to the Bäcklund transformation described in §5.4.2. (The important point about this – apparently arbitrary – separation of equation (5.95) is that the 'x' part is chosen to be analogous to the Sturm–Liouville equation as $u \to 0$. If we interpret D_x as $\partial/\partial x$ then the bilinear operator in equation (5.97) corresponds to

$(\partial^2/\partial x^2 + \lambda)$: see §5.2.1 and §4.2.) The connection with the Wahlquist & Estabrook (1973) representation is now obtained in a fairly straightforward manner.

First, we note that equation (5.97) can be written in the form

$$f_1 f_{2xx} + f_2 f_{1xx} - 2f_{1x}f_{2x} + \lambda f_1 f_2 = 0 \qquad (5.98)$$

(see §5.3.1). If we introduce

$$w_i = -2\frac{\partial}{\partial x}\log f_i = -2f_{ix}/f_i \qquad (i = 1, 2)$$

then equation (5.98) becomes, after a division by $\tfrac{1}{2}f_1 f_2$,

$$(w_1 + w_2)_x = 2\lambda + \tfrac{1}{2}(w_1 - w_2)^2, \qquad (5.99)$$

which is equation (5.80): the 'x' part of the Bäcklund transformation. Presumably, therefore, equation (5.96) is the 't' part of the transformation, and to confirm this we next write the equation as

$$f_1 f_{2t} - f_2 f_{1t} - 3\lambda(f_1 f_{2x} - f_2 f_{1x}) + f_1 f_{2xxx} - 3f_{1x}f_{2xx} + 3f_{2x}f_{1xx} - f_2 f_{1xxx} = 0,$$

by using the properties of the bilinear operator. If we now divide by $f_1 f_2$, differentiate once with respect to x and then introduce w_i, we obtain

$$-\tfrac{1}{2}(w_2 - w_1)_t + \tfrac{3}{2}\lambda(w_2 - w_1)_x + \tfrac{1}{2}(w_1 - w_2)_{xxx}$$
$$-\tfrac{3}{4}\{(w_1 - w_2)(w_1 + w_2)_x\}_x + \tfrac{1}{8}\{(w_1 - w_2)^3\}_x = 0.$$

Finally, if we eliminate λ by substituting from equation (5.99), the resulting equation can be simplified to

$$(w_1 - w_2)_t - 3(w_{1x}^2 - w_{2x}^2) + (w_1 - w_2)_{xxx} = 0$$

which is equation (5.81) – the 't' part of the Bäcklund transformation.

Further reading

5.1 The rôle of conservation laws, and the Lagrangian and Hamiltonian forms, in inverse scattering theory, are discussed by: Ablowitz & Segur (1981, §1.6); Calogero & Degasperis (1982, Chap. 5 and A.23). An introduction to Noether's theorem is given by Gel'fand & Fomin (1963, §§37, 38).

5.2 The Lax approach is described in Lax (1968); Calogero & Degasperis (1982, A.20).

5.3 Hirota's bilinear method is developed in detail in Matsuno (1984).

5.4 Bäcklund transformations are described in Rogers & Shadwick (1982); Lamb (1980, Chap. 8).

Exercises

Q5.1 *Burgers equation.* Show that, if $u(x, t)$ is a solution of

$$u_t + uu_x = vu_{xx}, \qquad -\infty < x < \infty,$$

where v is a (positive) constant, then u is a conserved density.

Q5.2 *KdV equation.* Show that $xu + 3tu^2$ is a conserved density for the KdV equation,

$$u_t - 6uu_x + u_{xxx} = 0.$$

Q5.3 *Modified KdV equation.* Find three conservation laws for the modified KdV equation

$$u_t - 6u^2 u_x + u_{xxx} = 0, \qquad -\infty < x < \infty,$$

which involve u, u^2 and u^4, respectively.

Q5.4 *Regularised long-wave equation.* The RLW equation can be written as

$$u_t - uu_x - u_{xxt} = 0, \qquad -\infty < x < \infty.$$

Show that u, $u^2 + u_x^2$ and u^3 are conserved densities. Note that X is here allowed to incorporate derivatives of u with respect to t.

[There are no other non-trivial independent conserved densities for the RLW equation: see Olver (1979). The *RLW equation*, sometimes called the *Peregrine equation* or the *BBM equation*, is discussed in Peregrine (1966) and Benjamin, Bona & Mahony (1972).]

Q5.5 *Nonlinear Schrödinger equation.* Show that, if $u(x, t)$ is a solution of the NLS equation

$$iu_t + u_{xx} + vu|u|^2 = 0, \qquad -\infty < x < \infty,$$

where v is a constant, then

$$\int_{-\infty}^{\infty} |u|^2 \, dx, \qquad \int_{-\infty}^{\infty} (u^* u_x - uu_x^*) \, dx \qquad \text{and} \qquad \int_{-\infty}^{\infty} (|u_x|^2 - \tfrac{1}{2}v|u|^4) dx$$

are constants of the motion. (The $*$ denotes the complex conjugate.)

Q5.6 *Benjamin–Ono equation.* Show that, if $u(x, t)$ is a solution of the Benjamin–Ono equation,

$$u_t + uu_x + \mathscr{H}(u_{xx}) = 0, \qquad -\infty < x < \infty,$$

where $\mathscr{H}$ is the Hilbert transform (see Q2.13), then

$$\int_{-\infty}^{\infty} u \, dx, \qquad \int_{-\infty}^{\infty} u^2 \, dx, \qquad \int_{-\infty}^{\infty} \{u^3 - 3u_x \mathscr{H}(u)\} \, dx \qquad \text{and} \qquad \frac{d}{dt} \int_{-\infty}^{\infty} xu \, dx$$

are independent of time.

Q5.7 *Sine–Gordon equation.* Given the sine–Gordon equation in the form

$$\phi_{xt} = \sin \phi,$$

verify the following conservation laws:

$$(\tfrac{1}{2}\phi_t^2)_x - (1 - \cos \phi)_t = 0 \qquad \text{and} \qquad (1 - \cos \phi)_x - (\tfrac{1}{2}\phi_x^2)_t = 0,$$
$$(\tfrac{1}{4}\phi_x^4 - \phi_{xx}^2)_t + (\phi_x^2 \cos \phi)_x = 0,$$
$$(\tfrac{1}{6}\phi_x^6 - \tfrac{2}{3}\phi_x^2\phi_{xx}^2 + \tfrac{8}{9}\phi_x^3\phi_{xxx} + \tfrac{4}{3}\phi_{xxx}^2)_t + \{(\tfrac{1}{9}\phi_x^4 - \tfrac{4}{3}\phi_{xx}^2)\cos \phi\}_x = 0.$$

(Lamb, 1971)

Q5.8 *Fréchet derivative.* The Fréchet (or variational) derivative, $\delta F/\delta u$, of the operator $F\{u\}$, is defined by

$$\int_{-\infty}^{\infty} v \frac{\delta F}{\delta u} \, dx = \lim_{\varepsilon \to 0} \frac{\partial}{\partial \varepsilon} \int_{-\infty}^{\infty} F(u + \varepsilon v) \, dx$$

for all continuous v. Show that, if $F(u) = f(u, u_x, u_{xx}, \ldots)$, then $\delta F/\delta u$ corresponds to the Euler–Lagrange operator, i.e.

$$\frac{\delta F}{\delta u} = \frac{\partial f}{\partial u} - \frac{d}{dx} \frac{\partial f}{\partial u_x} + \frac{d^2}{dx^2} \frac{\partial f}{\partial u_{xx}} \cdots.$$

Hence verify that

$$\frac{\partial u}{\partial t} = \frac{\partial}{\partial x} \left\{ \frac{\delta}{\delta u} (u^3 + \tfrac{1}{2} u_x^2) \right\}$$

is the KdV equation.

*Q5.9 *Conserved densities and the scattering problem.* Consider the classical scattering problem

$$\hat{\psi}_{xx} + (k^2 - u)\hat{\psi} = 0, \qquad -\infty < x < \infty,$$

with

$$\hat{\psi}(x; k) \sim e^{-ikx} + b(k)e^{ikx} \qquad \text{as } x \to +\infty,$$

for a given function $u(x)$. Introduce $h(x; k)$ such that

$$\hat{\psi}(x; k) = a(k)e^{-ikx + h}$$

for all real x and k, where a is the transmission coefficient, and deduce that

$$h_{xx} - 2ikh_x + h_x^2 - u = 0.$$

Now assume that

$$h_x(x; k) \sim \sum_{n=1}^{\infty} (2ik)^{-n} g_n(x) \qquad \text{as } k \to \infty$$

and determine

$$g_1 = -u; \qquad g_2 = -u_x; \qquad g_3 = -u_{xx} + u^2,$$

and, in general, that

$$g_n = \frac{d}{dx} g_{n-1} + \sum_{m=1}^{n-2} g_m g_{n-m-1}, \qquad n = 3, 4, \ldots.$$

Further show that, if k is now complex valued with $\mathscr{I}(k) > 0$, then

$$-\log a \sim \sum_{n=1}^{\infty} (2ik)^{-n} \int_{-\infty}^{\infty} g_n(x) \, dx \qquad \text{as } |k| \to \infty.$$

If u is a solution of the KdV equation, deduce that $\int_{-\infty}^{\infty} g_n \, dx$ is a constant of the motion for each $n = 1, 2, \ldots$.

[The asymptotic technique adopted here is the well-known WKBJ method; see Miura, Gardner & Kruskal (1968).]

Q5.10 *N-soliton solution.* Assume that $u(x, t)$ evolves, according to the KdV equation, into an N-soliton solution from a given initial profile, $u(x, 0)$. Consider the profile at $t = 0$, and the solution as $t \to \infty$, and hence show how the conserved quantities can be used to determine the amplitudes of the resulting solitons. Use the first two conservation laws, and then the first three, to verify your method for the two-soliton and three-soliton solutions, respectively.

(Berezin & Karpman, 1967)

Q5.11 *Unitarily equivalent operators.* Let L be a linear operator which acts on a Hilbert space H, and which depends on a parameter t. Define the adjoint, $\hat{U}$, of a linear operator U on H, so that the inner product $(u, Uv) = (\hat{U}u, v)$ for all $u, v \in H$. Show that, if $L(t)$ is *unitarily equivalent* to $L(0)$, i.e. if

$$\hat{U}(t)L(t)U(t) = L(0) \qquad \text{and} \qquad U(t)\hat{U}(t) = I,$$

where I is the identity operator, then $L(t)$ has the same eigenvalues as $L(0)$.

Assuming that such an operator U does exist, show that there exists an operator M on H such that

$$U_t = MU \qquad \text{where } \hat{M} = -M.$$

Hence deduce that, if L is self-adjoint, then

$$L_t + [L, M] = 0.$$

(Lax, 1968)

Q5.12 *Matrix Lax equation.* Show that, if $L(t)$ and $M(t)$ are $n \times n$ complex-valued matrices such that

$$L_t + [L, M] = 0,$$

where L is Hermitian and $\lambda(t)$ is a real eigenvalue of L, then λ is constant.

Q5.13 *An ordinary differential system.* Cast the system

$$\frac{dx}{dt} = gy, \qquad \frac{dy}{dt} = -gx,$$

where $g(x, y, t)$ is a given continuous function, into the equivalent form

$$L_t + [L, M] = 0,$$

finding L as some symmetric, and M as some antisymmetric, real matrix whose elements depend upon x, y and g (see Q5.12). Show that the eigenvalues of L are constant and hence deduce the (otherwise obvious) result that $x^2 + y^2$ is constant for each solution of the system.

Q5.14 *Two-dimensional KdV equation.* Show that the equation

$$(u_t - 6uu_x + u_{xxx})_x + 3u_{yy} = 0$$

can be obtained by choosing

$$L = -\frac{\partial^2}{\partial x^2} + \frac{\partial}{\partial y} + u; \qquad M = -4\frac{\partial^3}{\partial x^3} + 6u\frac{\partial}{\partial x} + 3u_x + 3\int^x u_y \, dx.$$

[Dryuma, 1974; this equation describes a class of waves with almost one-dimensional propagation (Johnson, 1980). This equation was introduced in Kadomtsev & Petviashvili (1970). A review of soliton interactions in two dimensions is given by Freeman (1980).]

Q5.15 *Hamiltonian form.* Show that the equation

$$-2\frac{\partial u}{\partial t} = \frac{\partial}{\partial x}\left(\frac{\delta T_4}{\delta u}\right)$$

(see Q5.8), where T_4 is given in §5.1.1, is the fifth-order KdV equation, (5.34). Also find the next equation in the KdV hierarchy which is obtained by replacing T_4 by T_5.

Q5.16 *Klein–Gordon equation.* Using the Lagrangian density

$$\mathscr{L} = \tfrac{1}{2}\{\psi_t^2 - (\nabla\psi)^2\} - V(\psi),$$

deduce the nonlinear Klein–Gordon equation

$$\psi_{tt} - \nabla^2\psi + V'(\psi) = 0$$

by requiring that the integral $\int \mathscr{L}\,dx\,dt$ is stationary with respect to small variations in ψ.

Find the Hamiltonian $\mathscr{H} = \psi_t \partial\mathscr{L}/\partial\psi_t - \mathscr{L}$. Deduce that the total 'energy', $\int \mathscr{H}\,dx$, can be written as

$$\int [\tfrac{1}{2}\{\psi_t^2 + (\nabla\psi)^2\} + V(\psi)]\,dx.$$

Q5.17 *Sine–Gordon equation.* Use the Lagrangian density given in Q5.16, with $V'(\psi) = \sin\psi$ and ∇ chosen as the one-dimensional operator $\partial/\partial x$, to obtain the sine–Gordon equation written in laboratory coordinates,

$$\psi_{xx} - \psi_{tt} = \sin\psi.$$

Q5.18 *Burgers equation.* Show that the solution of the Burgers equation

$$u_t + uu_x = \nu u_{xx},$$

where ν is a positive constant, can be reduced to the solution of the heat-conduction equation by use of the transformation

$$u(x, t) = A\frac{\partial}{\partial x}\log\theta(x, t)$$

for a suitable choice of the constant A. Hence show how to solve the Burgers equation.

[This equation has been used to model turbulent flow in a channel (Burgers, 1948), and also the structure of a shock wave (Lighthill, 1956). The transformation is sometimes attributed to Hopf and Cole: see Forsyth (1906, p. 101), Hopf (1950), Cole (1951).]

Q5.19 *Fifth-order KdV equations.* Introduce the transformation

$$u \to \alpha u, \qquad x \to \beta x, \qquad t \to \gamma t,$$

where α, β and γ are real constants, into one of the fifth-order equations,

(5.34) or (5.42), and deduce that it can not be transformed into the other.

Q5.20 *Bilinear operator.* Prove the following identities, where $D_t^m D_x^n(a \cdot b)$ is the bilinear operator defined by equation (5.37):

   (i) $D_t D_x(a \cdot 1) = D_t D_x(1 \cdot a) = \partial^2 a / \partial x \partial t$;

   (ii) $D_x^m(a \cdot b) = (-1)^m D_x^m(b \cdot a)$, and hence $D_x^m(a \cdot a) = 0$ for m odd;

   (iii) $D_t^m D_x^n(a \cdot b) = D_x^n D_t^m(a \cdot b)$;

   (iv) $D_t^m D_x^n \{\exp(\theta_1) \cdot \exp(\theta_2)\} = (\omega_2 - \omega_1)^m (k_1 - k_2)^n \exp(\theta_1 + \theta_2)$ where $\theta_i = k_i x - \omega_i t$, $i = 1, 2$;

   (v) $D_x(a \cdot b) = 0$ if $a \propto b$.

Q5.21 *Boussinesq equation.* Show that the bilinear form of the equation

$$u_{tt} - u_{xx} + 3(u^2)_{xx} - u_{xxxx} = 0$$

is

$$(D_t^2 - D_x^2 - D_x^4)(f \cdot f) = 0,$$

where $u(x, t) = -2(\partial^2 / \partial x^2) \log f$.

*Q5.22 *Benjamin–Ono equation.* Show that the bilinear form of the equation

$$u_t + 4uu_x + \mathcal{H}(u_{xx}) = 0,$$

where $\mathcal{H}$ is the Hilbert transform (see Q2.13), is

$$(iD_t - D_x^2)(f^* \cdot f) = 0,$$

where $u(x, t) = \frac{1}{2}i(\partial/\partial x) \log(f^*/f)$ and $*$ denotes the complex conjugate.

   [You may assume that $f(x, t) = \prod_{n=1}^{N} \{x - x_n(t)\}$, where $\mathcal{I}(x_n) > 0$ for $1 \leqslant n \leqslant N$.]

Q5.23 *Concentric KdV equation.* Show that the bilinear form of the equation

$$u_t + \frac{u}{2t} - 6uu_x + u_{xxx} = 0$$

is

$$\left(D_x D_t + D_x^4 + \frac{1}{2t}\frac{\partial}{\partial x}\right)(f \cdot f) = 0$$

where

$$u(x, t) = -2\frac{\partial^2}{\partial x^2}\log f \qquad \text{and} \qquad \frac{\partial}{\partial x}(a \cdot b) = \frac{\partial}{\partial x}\{a(x, t)b(x', t')\}$$

on $x' = x$, $t' = t$.

Q5.24 *Two-dimensional KdV equation.* Show that the bilinear form of the equation

$$(u_t - 6uu_x + u_{xxx})_x + 3u_{yy} = 0$$

is

$$(D_t D_x + D_x^4 + 3D_y^2)(f \cdot f) = 0$$

where $u = -2(\partial^2/\partial x^2) \log f$.

*Q5.25 *Sine–Gordon equation.* Show that the bilinear form of the equation

$$\phi_{xx} - \phi_{tt} = \sin \phi$$

is the pair

$$(D_x^2 - D_t^2 - 1)(f \cdot g) = 0$$
$$(D_x^2 - D_t^2)(f \cdot f - g \cdot g) = 0,$$

where $\phi = 4 \arctan(g/f)$.

*Q5.26 *Nonlinear envelope equation.* Show that the bilinear form of the equation

$$i\psi_t + 3i\alpha|\psi|^2\psi_x + \beta\psi_{xx} + i\gamma\psi_{xxx} + \delta\psi|\psi|^2 = 0,$$

where α, β, γ and δ are real constants such that $\alpha\beta = \gamma\delta$, is the pair

$$(iD_t + \beta D_x^2 + i\gamma D_x^3)(g \cdot f) = 0,$$
$$\gamma D_x^2(f \cdot f) = \alpha g g^*,$$

where $\psi = g/f$, f is real and an asterisk denotes the complex conjugate.
(Hirota, 1973a)

*Q5.27 *Fifth-order KdV equation* (*Lax hierarchy*). Introduce an auxiliary independent variable, τ, such that

$$u_\tau - 6uu_x + u_{xxx} = 0,$$

and hence show that the bilinear form of the equation

$$u_t + 30u^2u_x - 20u_xu_{xx} - 10uu_{xxx} + u_{xxxxx} = 0$$

is given by the pair

$$\{D_x(D_t + D_x^5) - \tfrac{5}{6}D_x^3(D_\tau + D_x^3)\}(f \cdot f) = 0,$$
$$D_x(D_\tau + D_x^3)(f \cdot f) = 0,$$

where $u = -2(\partial^2/\partial x^2)\log f$.

Q5.28 *Solitary-wave solutions.* Obtain the solitary-wave solutions of the equations given in Q5.21–Q5.27 by seeking simple solutions of the corresponding bilinear form.

[In the case of Q5.27, τ appears in the arbitrary phase shift in the exponent; the exponent is linear in x, t and τ.]

Q5.29 *Three-soliton solution.* Obtain the solution for $f(x, t)$ which represents the three-soliton solution of the KdV equation, where f satisfies the bilinear form

$$D_x(D_t + D_x^3)(f \cdot f) = 0.$$

Q5.30 *Two-dimensional KdV equation: two-soliton solution.* Use the bilinear form (Q5.24) of the 2D KdV equation,

$$(u_t - 6uu_x + u_{xxx})_x + 3u_{yy} = 0,$$

to obtain the two-soliton solution as

$$f = 1 + E_1 + E_2 + A_2E_1E_2,$$

where $E_i = \exp(\theta_i)$, $\theta_i = k_ix + l_iy - \omega_it + \alpha_i$, $\omega_i = k_i^3 + 3l_i^2/k_i$, for $i = 1, 2$ and A_2 is suitably chosen.

Discuss the behaviour of the solution, u, for θ_1 fixed, as $\theta_2 \to \pm\infty$; and for θ_2 fixed, as $\theta_1 \to \pm\infty$.

Q5.31 *Liouville's equation.* By using the Bäcklund transformation for Liouville's equation, and the equation $v_{xt} = 0$ (see §5.4.1), obtain the general solution of Liouville's equation.

Q5.32 *Burgers equation.* Use the Bäcklund transformation

$$v_x = -uv/2v, \qquad v_t = (u^2 - 2vu_x)v/4v$$

to show that

$$u_t + uu_x = vu_{xx} \qquad \text{and} \qquad v_t = vv_{xx}.$$

Q5.33 *Sine–Gordon equation.* Use the Bäcklund transformation for the sine–Gordon equation (see equations (5.68)) to show that four solutions u_i ($i = 1, \ldots, 4$) are related by

$$\tan\left(\frac{u_4 - u_1}{4}\right) = \frac{a_1 + a_2}{a_1 - a_2} \tan\left(\frac{u_2 - u_3}{4}\right),$$

where a_1, a_2 are non-zero constants.

Q5.34 *Sine–Gordon equation: soliton solutions.* Use the relation given in Q5.33 to obtain a soliton solution of the sine–Gordon equation (for example, by choosing $u_1 = 0$ and u_2 and u_3 different solitary-wave solutions).

*Q5.35 *KdV equation.* Use the algebraic form of the Bäcklund transformation to obtain the three-soliton solution of the KdV equation.

Q5.36 *Some exchange formulae.* Verify these identities for the bilinear operator:

   (i) $ggD_xD_t(f \cdot f) - ffD_x^2(g \cdot g) = 2D_x[\{D_t(f \cdot g)\} \cdot (fg)]$;

   (ii) $ggD_x^2(f \cdot f) - ffD_x^2(g \cdot g) = 2D_x[\{D_x(f \cdot g)\} \cdot (fg)]$;

   (iii) $ggD_x^4(f \cdot f) - ffD_x^4(g \cdot g) = 2D_x[\{D_x^3(f \cdot g)\} \cdot (fg)]$
   $$+ 6D_x[\{D_x^2(f \cdot g)\} \cdot \{D_x(g \cdot f)\}]$$
   $$= 2D_x^3[\{D_x(f \cdot g)\} \cdot (fg)];$$

   (iv) $D_x[\{D_x(f \cdot g)\} \cdot \{D_x(g \cdot f)\}] = 0.$

*Q5.37 *Boussinesq equation.* Show, from the bilinear form, that a Bäcklund transformation for the equation

$$u_{tt} - u_{xx} + 3(u^2)_{xx} - u_{xxxx} = 0$$

is

$$(D_t + aD_x^2)(f_1 \cdot f_2) = 0$$
$$(aD_tD_x + D_x + D_x^3)(f_1 \cdot f_2) = 0,$$

where $u(x, t) = -2(\partial^2/\partial x^2)\log f$ and $a^2 = -3$; see Q5.21.

[The term $\lambda f_1 f_2$, for arbitrary λ, may be inserted on the right-hand side of the second equation in the Bäcklund transformation to give a more general transformation.]

*Q5.38 *Sawada–Kotera equation.* Show, from the bilinear form, that a Bäcklund transformation for the equation

$$u_t + 45u^2u_x - 15u_xu_{xx} - 15uu_{xxx} + u_{xxxxx} = 0$$

is

$$D_x^3(f_1 \cdot f_2) = \lambda f_1 f_2$$
$$(D_t - \tfrac{15}{2}\lambda D_x^2 - \tfrac{3}{2}D_x^5)(f_1 \cdot f_2) = 0,$$

where λ is an arbitrary constant and $u(x,t) = -2(\partial^2/\partial x^2)\log f$; see equation (5.41).

Q5.39 *An NLS equation.* Verify that an auto-Bäcklund transformation for the NLS equation

$$iu_t + u_{xx} + 2u|u|^2 = 0$$

is the pair

$$u_x + v_x = (u-v)(4\lambda^2 - |u+v|^2)^{1/2}, \tag{1}$$

$$u_t + v_t = i(u_x - v_x)(4\lambda^2 - |u+v|^2)^{1/2} + \tfrac{1}{2}i(u+v)(|u+v|^2 + |u-v|^2), \tag{2}$$

where both u and v satisfy the equation. Hence obtain the solitary-wave solution of this NLS equation (cf. Q2.3) by choosing $v = 0$ for all x, t.

(Konno & Wadati, 1975)

Q5.40 *A modified KdV equation.* Let $u = w_x$, and define the operators

$$P(u) = u_t + 6u^2 u_x + u_{xxx}; \qquad Q(w) = w_t + 2w_x^3 + w_{xxx}.$$

Show that $P(u) = \{Q(w)\}_x$.

Use the Bäcklund transformation

$$w_{2x} = -w_{1x} + 2k\sin(w_2 - w_1)$$

$$w_{2t} = -w_{1t} - 2k\{(u_{2x} - u_{1x})\cos(w_2 - w_1) + (u_2^2 - u_1^2)\sin(w_2 - w_1)\},$$

for $u = u_i$, $w = w_i$ $(i = 1, 2)$, where k is an arbitrary constant, to show that $Q(w_i) = 0$ and so $P(u_i) = 0$.

If $w_1 = 0$, for all x and t, show that

$$w_2(x,t) = \pm 2\arctan[\exp\{2k(x - x_0 - 4k^2 t)\}],$$

where x_0 is another arbitrary constant.

Now, given w_0, k_1 and k_2, introduce the functions w_1, w_2, w_{12} and w_{21} such that

$$(w_1 + w_0)_x = 2k_1\sin(w_1 - w_0); \qquad (w_2 + w_0)_x = 2k_2\sin(w_2 - w_0);$$

$$(w_{12} + w_1)_x = 2k_2\sin(w_{12} - w_1); \qquad (w_{21} + w_2)_x = 2k_1\sin(w_{21} - w_2).$$

Use the theorem of permutability (i.e. $w_{12} = w_{21}$) to deduce that

$$w_{12} = w_0 + 2\arctan\left(\frac{(k_1 + k_2)}{(k_1 - k_2)}\tan\tfrac{1}{2}(w_2 - w_1)\right).$$

(Wadati, 1974; Lamb, 1980, p. 257)

More general inverse methods

In Chap. 4 we discussed the inverse scattering transform for the KdV equation in detail. Now this novel procedure would be even more useful if it could be extended or generalised to accommodate other equations. Indeed, we have already hinted that this is possible. In this chapter we shall describe two approaches to this generalisation, but which are quite closely related.

In 1972 (1971 in Russian), Zakharov & Shabat (ZS) published the inverse scattering transform for the nonlinear Schrödinger (NLS) equation,

$$iu_t + u_{xx} + u|u|^2 = 0.$$

Shortly thereafter they extended the technique to other equations (Shabat, 1973, Zakharov & Shabat, 1974). This extension, which we shall call the *ZS scheme*, essentially takes the Lax method and recasts it in a matrix form (see Q5.12 and Q5.13), leading directly to a matrix Marchenko equation.

At about the same time, a group at Clarkson College (in Potsdam, New York State) were developing an equivalent scheme which generalises the method that we have described here for the KdV equation (and which is close to the early work of Zakharov & Shabat). This culminated in the paper by Ablowitz, Kaup, Newell & Segur (AKNS), published in 1974, which immediately enabled the inverse scattering transform to be applied to many evolution equations. The *AKNS scheme* starts from a generalisation of the Sturm–Liouville equation by regarding it as a pair of first order equations. (This is usually referred to as a *2 × 2 eigenvalue problem*.)

Both the AKNS and ZS schemes are applicable to the KdV and NLS equations as well as, for example, to the modified KdV (mKdV) and sine–Gordon (SG) equations. (In particular we shall describe, later in this chapter, some properties of both the NLS and SG equations.) It is no surprise that the two schemes overlap, but the important distinction is that the AKNS scheme is couched in terms of scattering theory whereas the ZS scheme (in its final form) is expressed solely in terms of operators.

We shall describe the main ingredients in the AKNS scheme, although we shall stop short of presenting all the technical details. We shall also outline the ZS scheme and briefly mention how the two are related.

6.1 The AKNS scheme

6.1.1 The 2 × 2 eigenvalue problem

We begin this discussion by introducing the pair of equations

$$\psi_{1x} = -i\zeta\psi_1 + q\psi_2, \tag{6.1}$$

$$\psi_{2x} = i\zeta\psi_2 + r\psi_1, \tag{6.2}$$

i.e.

$$\psi_x = \begin{pmatrix} -i\zeta & q \\ r & i\zeta \end{pmatrix}\psi, \quad \text{or} \quad \begin{pmatrix} \partial/\partial x & -q \\ r & -\partial/\partial x \end{pmatrix}\psi = -i\zeta\psi, \tag{6.3a, b}$$

where ψ is the two-vector $\begin{pmatrix} \psi_1 \\ \psi_2 \end{pmatrix}$. The (bounded) functions $q(x)$ and $r(x)$ – not necessarily real – are *potentials* and ζ is the *eigenvalue*. It is now straightforward to demonstrate the connection between the pair of equations (6.1) and (6.2), and our original Sturm–Liouville equation. First, the differential of equation (6.2) with respect to x is

$$\psi_{2xx} = i\zeta\psi_{2x} + r_x\psi_1 + r\psi_{1x}$$

(provided r_x exists). Then, using equations (6.1) and (6.2), we obtain

$$\psi_{2xx} = i\zeta\psi_{2x} + r_x\psi_1 + r(-i\zeta\psi_1 + q\psi_2)$$
$$= i\zeta\psi_{2x} + (r_x - i\zeta r)(\psi_{2x} - i\zeta\psi_2)/r + qr\psi_2$$

or

$$\psi_{2xx} - \frac{r_x}{r}\psi_{2x} - \left(qr - i\zeta\frac{r_x}{r} - \zeta^2\right)\psi_2 = 0.$$

The special choice $r = -1$ now gives the Sturm–Liouville equation for ψ_2,

$$\psi_{2xx} + (\zeta^2 + q)\psi_2 = 0;$$

see equation (3.1) with $\lambda = \zeta^2$ and $u = -q$. Thus the system, given by equations (6.3), recovers the scattering equation required for the solution of the KdV equation. It should be noted, however, that the choice $r = -1$ turns out to be a degenerate case since hereafter, for all other evolution equations, we assume that both $q(x)$ and $r(x)$ decay sufficiently rapidly as $|x| \to \infty$. This ensures the existence of ψ for $x\in(-\infty, \infty)$.

We now introduce four solutions of equations (6.3), which are defined by their asymptotic behaviours at infinity

$$\psi_+(x;\zeta) \sim e^{-i\zeta x}\begin{pmatrix}1\\0\end{pmatrix}; \qquad \hat{\psi}_+(x;\zeta) \sim e^{i\zeta x}\begin{pmatrix}0\\1\end{pmatrix}, \qquad \text{as } x \to +\infty \quad (6.4)$$

and

$$\psi_-(x;\zeta) \sim e^{-i\zeta x}\begin{pmatrix}1\\0\end{pmatrix}; \qquad \hat{\psi}_-(x;\zeta) \sim e^{i\zeta x}\begin{pmatrix}0\\-1\end{pmatrix}, \qquad \text{as } x \to -\infty. \quad (6.5)$$

These solutions are written more compactly, in matrix notation, as

$$\Psi_+ = \begin{pmatrix}\psi_{1+} & \hat{\psi}_{1+}\\\psi_{2+} & \hat{\psi}_{2+}\end{pmatrix}; \qquad \Psi_- = \begin{pmatrix}\psi_{1-} & \hat{\psi}_{1-}\\\psi_{2-} & \hat{\psi}_{2-}\end{pmatrix}. \quad (6.6)$$

These are linearly dependent, and may be connected via the *scattering matrix*, $S(\zeta)$, by

$$\Psi_- = \Psi_+ S \qquad \text{where } S = \begin{pmatrix}a & \hat{b}\\b & -\hat{a}\end{pmatrix}. \quad (6.7)$$

The scattering coefficients $a, b, \hat{a}$ and $\hat{b}$ are functions of ζ, and the two-vectors $\psi_+, \hat{\psi}_+, \psi_-$ and $\hat{\psi}_-$ are, according to equation (6.7), related by

$$\psi_- = a\psi_+ + b\hat{\psi}_+; \qquad \hat{\psi}_- = \hat{b}\psi_+ - \hat{a}\hat{\psi}_+. \quad (6.8)$$

(It should be noted that the scattering coefficients here do not correspond directly with those introduced for the solution of the KdV equation: see equation (3.6). The correspondence requires, for example, $a^{-1} \to a$ and $ba^{-1} \to b$.) The problem described by equation (6.3), with (6.4), (6.5) and (6.7), is the *2 × 2 eigenvalue problem*.

The solution of the pair of first order equations, (6.1) and (6.2), can always be represented by an integral over an appropriate kernel (cf. equation (3.28)). Thus we assume that the solution for Ψ_+ can be expressed as

$$\Psi_+(x;\zeta) = \Psi_0(x;\zeta) + \int_x^\infty K(x,y)\Psi_0(y;\zeta)\,\mathrm{d}y \quad (6.9)$$

where K is an appropriate 2×2 matrix with elements K_{ij} $(i,j = 1,2)$, and

$$\Psi_0(x;\zeta) = \begin{pmatrix}e^{-i\zeta x} & 0\\0 & e^{i\zeta x}\end{pmatrix}, \quad (6.10)$$

which describes the behaviours given in (6.4), so that $\Psi_+ \to \Psi_0$ as $x \to +\infty$. (It is obvious that we can obtain Ψ_- from equation (6.7), once Ψ_+ has been determined.) We note that equation (6.9) defines Ψ_+ in terms of a kernel, K, which is not a function of the eigenvalue, ζ. That such a kernel exists will now be demonstrated.

In order to find the equations satisfied by the elements of K, it is necessary to ensure that the two columns of Ψ_+ are indeed solutions of equation (6.3a or b). This is most easily accomplished by operating on equation (6.9) with

$$\begin{pmatrix} \partial/\partial x & -q \\ r & -\partial/\partial x \end{pmatrix}.$$

Immediately, from equations (6.3b) and (6.9), we see that

$$-i\zeta\Psi_+ = -i\zeta\left(\Psi_0 + \int_x^\infty K\Psi_0\,dy\right), \tag{6.11}$$

(where we have suppressed the arguments, for simplicity). The first term on the right in equation (6.9) also gives

$$\begin{pmatrix} \partial/\partial x & -q \\ r & -\partial/\partial x \end{pmatrix}\Psi_0 = \begin{pmatrix} -i\zeta & -q \\ r & -i\zeta \end{pmatrix}\Psi_0 = -i\zeta\Psi_0 + Q\Psi_0, \tag{6.12}$$

where Q is the matrix $\begin{pmatrix} 0 & -q \\ r & 0 \end{pmatrix}$. Finally, we operate on the integral in equation (6.9); this gives

$$\begin{pmatrix} \partial/\partial x & -q \\ r & -\partial/\partial x \end{pmatrix}\int_x^\infty \begin{pmatrix} K_{11}e^{-i\zeta y} & K_{12}e^{i\zeta y} \\ K_{21}e^{-i\zeta y} & K_{22}e^{i\zeta y} \end{pmatrix}dy. \tag{6.13}$$

We can now find, for example, the first element $(i=j=1)$ in the representation of expression (6.13): this is

$$\frac{\partial}{\partial x}\int_x^\infty K_{11}(x,y)e^{-i\zeta y}\,dy - q(x)\int_x^\infty K_{21}(x,y)e^{-i\zeta y}\,dy$$

$$= -K_{11}(x,x)e^{-i\zeta x} + \int_x^\infty K_{11x}e^{-i\zeta y}\,dy - q(x)\int_x^\infty K_{21}e^{i\zeta y}\,dy,$$

if $K_{11}(x,y)\to 0$ as $y\to\infty$; and similar expressions are obtained for the other three elements. Collecting these together, we see that the matrix (6.13) becomes

$$\hat{I}K(x,x)\Psi_0 - \hat{I}\int_x^\infty K_x\Psi_0\,dy + Q\int_x^\infty K\Psi_0\,dy$$

and so operating on equation (6.9) transforms it into

$$-i\zeta\left(\Psi_0 + \int_x^\infty K\Psi_0\,dy\right) = -(i\zeta I - \hat{I}K)\Psi_0 - \hat{I}\int_x^\infty K_x\Psi_0\,dy$$

$$+ Q\left(\Psi_0 + \int_x^\infty K\Psi_0\,dy\right), \tag{6.14}$$

where I is the unit matrix, and $\hat{I} = \begin{pmatrix} -1 & 0 \\ 0 & 1 \end{pmatrix}$. (Observe that $\hat{I}^2 = I$.)

To cast equation (6.14) into its most useful form we note that, if $K(x, y) \to 0$ as $y \to \infty$, then

$$-i\zeta \int_x^\infty K\Psi_0 \, dy = K(x, x)\Psi_0 \hat{I} + \int_x^\infty K_y \Psi_0 \hat{I} \, dy,$$

after an integration by parts. Thus equation (6.14) can be rewritten as

$$\hat{I}K(x, x)\Psi_0(x; \zeta) - K(x, x)\Psi_0(x; \zeta)\hat{I} + Q(x)\Psi_0(x; \zeta)$$

$$- \int_x^\infty \{\hat{I}K_x(x, y)\Psi_0(y; \zeta) + K_y(x, y)\Psi_0(y; \zeta)\hat{I} - Q(x)K(x, y)\Psi_0(y; \zeta)\} \, dy = 0$$

which is satisfied if $K(x, y)$ is a solution of

$$\hat{I}K_x + K_y\hat{I} - Q(x)K = 0, \tag{6.15}$$

and

$$\hat{I}K(x, x) - K(x, x)\hat{I} + Q(x) = 0. \tag{6.16}$$

(Note that $K\Psi_0\hat{I} = K\hat{I}\Psi_0$.) From equation (6.15) we obtain the set of four scalar equations,

$$\begin{aligned} K_{11x} + K_{11y} - qK_{21} = 0; && K_{22x} + K_{22y} - rK_{12} = 0; \\ K_{12x} - K_{12y} - qK_{22} = 0; && K_{21x} - K_{21y} - rK_{11} = 0, \end{aligned} \tag{6.17}$$

and from equation (6.16) the conditions

$$K_{12}(x, x) = -\tfrac{1}{2}q(x); \qquad K_{21}(x, x) = -\tfrac{1}{2}r(x). \tag{6.18}$$

(These equations are a generalisation of those used for the KdV equation: see equations (3.29) and (3.30), and Q4.1.) It is clear that $K(x, y)$ is not a function of the eigenvalue (ζ) and, furthermore, it is now possible to show that the solution for K exists and is unique. In other words the integral representation, (6.9), is defined.

As in the discussion for the KdV equation, once the potentials q and r are given, the scattering problem may be solved (at least in principle) and the scattering matrix determined. On the other hand, in the context of solving evolution equations, we know that the procedure involves solving the inverse problem. This 2×2 problem is no exception: that is, given the scattering matrix $S(\zeta)$, we shall need to know how to reconstruct the potentials $q(x)$ and $r(x)$. It is this aspect which we shall now examine.

6.1.2 The inverse scattering problem

The first stage in this development involves rewriting equations (6.8) as

$$\hat{\Psi}_- = \Psi_+ T, \tag{6.19}$$

where

$$\hat{\Psi}_- = \begin{pmatrix} \psi_{1-}/a & \hat{\psi}_{1-}/\hat{a} \\ \psi_{2-}/a & \hat{\psi}_{2-}/\hat{a} \end{pmatrix} \quad \text{and} \quad T(\zeta) = \begin{pmatrix} 1 & \hat{b}/\hat{a} \\ b/a & -1 \end{pmatrix}.$$

From equation (6.9) we now substitute for Ψ_+ into equation (6.19) to give

$$\hat{\Psi}_- = \left(\Psi_0 + \int_x^\infty K \Psi_0 \, dy \right) T, \tag{6.20}$$

where again we have suppressed the arguments. In order to derive an integral equation for K, given $T(\zeta)$, it is convenient to post-multiply equation (6.20) by $\frac{1}{2}\pi^{-1} \Psi_0(z;\zeta)$, for $z > x$, and then integrate along an appropriate contour in the complex ζ-plane from $-\infty + i0$ to $+\infty + i0$. This contour is indented into the upper half-plane for terms involving $e^{i\zeta z}$, and into the lower half-plane for $e^{-i\zeta z}$; we shall call these contours C_+ and C_-, respectively. (We shall write more about them later.) Thus equation (6.20) becomes

$$\frac{1}{2\pi} \int_C \hat{\Psi}_-(x;\zeta) \Psi_0(z;\zeta) \, d\zeta = \frac{1}{2\pi} \int_C \Psi_0(x;\zeta) T(\zeta) \Psi_0(z;\zeta) \, d\zeta$$
$$+ \frac{1}{2\pi} \int_C \left\{ \int_x^\infty K(x,y) \Psi_0(y;\zeta) \, dy \right\} T(\zeta) \Psi_0(z;\zeta) \, d\zeta, \tag{6.21}$$

where the two contours are represented by C, to allow the equation to be written in a compact form. (This is analogous to equation (3.33).)

The term involving the double integral in equation (6.21) is expressed, by changing the order of integration, as

$$\frac{1}{2\pi} \int_x^\infty K(x,y) \left\{ \int_C \Psi_0(y;\zeta) T(\zeta) \Psi_0(z;\zeta) \, d\zeta \right\} dy.$$

Further, we note that

$$\Psi_0(x;\zeta) T(\zeta) \Psi_0(z;\zeta) = \begin{pmatrix} E_- & (\hat{b}/\hat{a})E_+^* \\ (b/a)E_+ & -E_-^* \end{pmatrix},$$

where $E_\pm = e^{i\zeta(z \pm x)}$ and an asterisk denotes the complex conjugate. It is now convenient to define

$$f(X) = \frac{1}{2\pi} \int_{C_+} \frac{b(\zeta)}{a(\zeta)} e^{i\zeta x} \, d\zeta; \qquad \hat{f}(X) = \frac{1}{2\pi} \int_{C_-} \frac{\hat{b}(\zeta)}{\hat{a}(\zeta)} e^{-i\zeta x} \, d\zeta, \tag{6.22}$$

where C_+ goes *above* all the zeros of $a(\zeta)$ in the upper half-plane, and C_- *below* all those of $\hat{a}(\zeta)$ in the lower half-plane. Since we also have

$$\frac{1}{2\pi} \int_C e^{i\zeta(z-x)} \, d\zeta = \delta(z-x),$$

where δ is the Dirac delta function, we see that

$$\frac{1}{2\pi} \int_C \Psi_0(x;\zeta) T(\zeta) \Psi_0(z;\zeta) \, d\zeta = \begin{pmatrix} \delta(z-x) & \hat{f}(x+z) \\ f(x+z) & -\delta(z-x) \end{pmatrix}.$$

Thus equation (6.21) becomes

$$\frac{1}{2\pi}\int_C \hat{\Psi}_-(x;\zeta)\Psi_0(z;\zeta)\,d\zeta = \hat{F}(x+z) - K(x,z)\hat{I} + \int_x^\infty K(x,y)\hat{F}(y+z)\,dy,$$

(6.23)

for $z > x$, where the matrix $\hat{F} = \begin{pmatrix} 0 & \hat{f} \\ \hat{f} & 0 \end{pmatrix}$.

Finally, by analogy with the argument given in §3.3, it can be shown that, with C_+ and C_- as defined earlier,

$$\int_C \hat{\Psi}_-(x;\zeta)\Psi_0(z;\zeta)\,d\zeta = 0,$$

and so equation (6.23) becomes

$$\hat{F}(x+z) - K(x,z)\hat{I} + \int_x^\infty K(x,y)\hat{F}(y+z)\,dy = 0.$$

This is the *matrix Marchenko equation* for $K(x,z)$ which can be written in the more conventional form

$$K(x,z) + F(x+z) + \int_x^\infty K(x,y)F(y+z)\,dy = 0,$$

(6.24)

where $\hat{F} = -F\hat{I}$, i.e. $F = \begin{pmatrix} 0 & -\hat{f} \\ f & 0 \end{pmatrix}$; cf. equation (3.37) for the KdV problem. The solution of this system of equations yields $K(x,z)$ and then the potentials are given by

$$q(x) = -2K_{12}(x,x); \qquad r(x) = -2K_{21}(x,x);$$

(6.25)

see equations (6.18).

In deriving the Marchenko equation in Chap. 3, we found it convenient to deal separately with the discrete and the continuous spectra. For the 2×2 eigenvalue problem discussed here, the two corresponding contributions are already implied by the definitions (6.22) of f and $\hat{f}$. To see this, let us suppose that $a(\zeta)$ has N simple zeros at $\zeta = \zeta_n$, $n = 1, 2, \ldots, N$, in the upper half-plane. An application of Cauchy's residue theorem to the closed contour comprising C_+ and the real axis gives

$$f(X) = \frac{1}{2\pi}\int_{-\infty}^\infty \frac{b(k)}{a(k)}\exp(ikX)\,dk - i\sum_{n=1}^N c_n \exp(i\zeta_n X),$$

(6.26)

where $c_n = b/(da/d\zeta)$ evaluated at $\zeta = \zeta_n$. The integral here is now along the real axis only. (Alternatively, we can find c_n from equations (6.8) evaluated at $\zeta = \zeta_n$, so that

$$\psi_{i-} = b\hat{\psi}_{i+} \quad \text{for} \quad i = 1, 2.$$

This allows the determination of $b(\zeta_n)$ directly, a useful manoeuvre if $b(\zeta) = 0$ for all $\zeta \neq \zeta_n$.) Similarly, if $\hat{a}(\zeta)$ has M simple zeros at $\zeta = \zeta_m, m = 1, 2, \ldots, M$, in the lower half-plane, then

$$\hat{f}(X) = \frac{1}{2\pi} \int_{-\infty}^{\infty} \frac{\hat{b}(k)}{\hat{a}(k)} \exp(-ikX)\,dk + i \sum_{m=1}^{M} \hat{c}_m \exp(-i\zeta_m X), \quad (6.27)$$

where $\hat{c}_m = \hat{b}/(d\hat{a}/d\zeta)$ at $\zeta = \zeta_m$. By analogy with the classical Schrödinger scattering problem, the zeros of $a(\zeta)$ are called the *discrete eigenvalues*, the c_n are the *normalisation constants* and b/a is the *reflection coefficient*. Thus, if $b/a = 0$ for all real ζ, the scattering is produced by a *reflectionless potential*. Corresponding interpretations for the lower half-plane are usually redundant since we are normally concerned with problems where there is a relation between q and r, e.g. $r = -q$ or $r = q^*$. In such problems a symmetry exists between a and $\hat{a}$, c_n and $\hat{c}_m$, and so on; some of these points are discussed in §6.1.5 and also in the exercises at the end of the chapter.

6.1.3 An example: $r = -q, q = \lambda$ sech λx

In order to clarify some of the ideas involved in the 2×2 eigenvalue problem, and its inverse, we shall describe this simple example. First, we suppose that the scattering problem is defined with $r = -q$ (which will turn out to be relevant to the solution of the modified KdV equation (see Q6.7) and the sine–Gordon equation (see §6.1.5(b)).) In particular we choose the potential

$$q(x) = \lambda \operatorname{sech} \lambda x,$$

where λ is a positive constant. We therefore require the solution of

$$\begin{pmatrix} \partial/\partial x & -q \\ -q & -\partial/\partial x \end{pmatrix} \psi = -i\zeta\psi \qquad (6.28)$$

(see equation (6.3b)), for which it is convenient to write $\psi = E\phi$ where $E = e^{\pm i\zeta x}$ and ϕ is the two-vector $\begin{pmatrix} \phi_1 \\ \phi_2 \end{pmatrix}$. After substitution into equation (6.28) it follows that

$$\begin{pmatrix} \partial/\partial x & -q \\ -q & -\partial/\partial x \end{pmatrix} \phi = -2i\zeta\theta_{\pm}, \qquad (6.29)$$

where $\theta_+ = \begin{pmatrix} \phi_1 \\ 0 \end{pmatrix}$ if $E = e^{+i\zeta x}$, and $\theta_- = \begin{pmatrix} 0 \\ \phi_2 \end{pmatrix}$ if $E = e^{-i\zeta x}$.

The solution of equation (6.29), with $q(x) = \lambda$ sech λx, is surprisingly straightforward: in the case when $E = e^{+i\zeta x}$ suppose that $\phi_1 = A$ sech λx

where A is an arbitrary constant, and then

$$\phi_{2x} = -\lambda A \operatorname{sech}^2 \lambda x.$$

Thus

$$\phi_2 = -A \tanh \lambda x + B,$$

where B is a second arbitrary constant and, since

$$\phi_{1x} - \lambda \phi_2 \operatorname{sech} \lambda x = -2\mathrm{i}\zeta \phi_1,$$

we require that $B = 2\mathrm{i}\zeta A/\lambda$. Hence one solution of equation (6.28) is

$$\psi = \frac{A}{\lambda} \mathrm{e}^{\mathrm{i}\zeta x} \begin{pmatrix} \lambda s \\ 2\mathrm{i}\zeta - \lambda t \end{pmatrix},$$

where $s(x) = \operatorname{sech} \lambda x$ and $t(x) = \tanh \lambda x$. Now, since $s \to 0$ as $x \to \pm \infty$, we may choose A in one of two ways: (i) if $\psi_2 \sim \mathrm{e}^{\mathrm{i}\zeta x}$ as $x \to +\infty$ then $A(2\mathrm{i}\zeta - \lambda)/\lambda = 1$; (ii) if $\psi_2 \sim -\mathrm{e}^{\mathrm{i}\zeta x}$ as $x \to -\infty$ then $A(2\mathrm{i}\zeta + \lambda)/\lambda = -1$. Thus we have obtained the pair of solutions

$$\hat{\psi}_+ = \frac{\mathrm{e}^{\mathrm{i}\zeta x}}{2\mathrm{i}\zeta - \lambda} \begin{pmatrix} \lambda s \\ 2\mathrm{i}\zeta - \lambda t \end{pmatrix}; \qquad \hat{\psi}_- = \frac{-\mathrm{e}^{\mathrm{i}\zeta x}}{2\mathrm{i}\zeta + \lambda} \begin{pmatrix} \lambda s \\ 2\mathrm{i}\zeta - \lambda t \end{pmatrix}, \qquad (6.30)$$

(see equations (6.4) and (6.5)). Similarly, if we let $\phi_2 = A \operatorname{sech} \lambda x$ when $E = \mathrm{e}^{-\mathrm{i}\zeta x}$ with θ_-, we find the pair of solutions

$$\psi_+ = \frac{\mathrm{e}^{-\mathrm{i}\zeta x}}{2\mathrm{i}\zeta + \lambda} \begin{pmatrix} 2\mathrm{i}\zeta + \lambda t \\ \lambda s \end{pmatrix}; \qquad \psi_- = \frac{\mathrm{e}^{-\mathrm{i}\zeta x}}{2\mathrm{i}\zeta - \lambda} \begin{pmatrix} 2\mathrm{i}\zeta + \lambda t \\ \lambda s \end{pmatrix}. \qquad (6.31)$$

The four solutions given in (6.30) and (6.31) are connected, via the scattering matrix $S(\zeta)$, according to equations (6.7) and (6.8). Thus, for example,

$$\frac{\mathrm{e}^{-\mathrm{i}\zeta x}}{2\mathrm{i}\zeta - \lambda} \begin{pmatrix} 2\mathrm{i}\zeta + \lambda t \\ \lambda s \end{pmatrix} = \frac{a\mathrm{e}^{-\mathrm{i}\zeta x}}{2\mathrm{i}\zeta + \lambda} \begin{pmatrix} 2\mathrm{i}\zeta + \lambda t \\ \lambda s \end{pmatrix} + \frac{b\mathrm{e}^{\mathrm{i}\zeta x}}{2\mathrm{i}\zeta - \lambda} \begin{pmatrix} \lambda s \\ 2\mathrm{i}\zeta - \lambda t \end{pmatrix} \qquad (6.32)$$

for all x, which is possible only if

$$a(\zeta) = \frac{2\mathrm{i}\zeta + \lambda}{2\mathrm{i}\zeta - \lambda} \qquad \text{and} \qquad b(\zeta) = 0 \ (\text{if } a \neq 0). \qquad (6.33)$$

Similarly, we can show that

$$\hat{a}(\zeta) = \frac{2\mathrm{i}\zeta - \lambda}{2\mathrm{i}\zeta + \lambda} \qquad \text{and} \qquad \hat{b}(\zeta) = 0 \ (\text{if } \hat{a} \neq 0), \qquad (6.34)$$

so that in this case, where $r = -q$ and $q(x) = \lambda \operatorname{sech} \lambda x$, we find that $\hat{a} = a^*$ and the potential is reflectionless (i.e. $b = \hat{b} = 0$ for all real ζ). Furthermore, we observe that $a(\zeta)$ has one zero in the upper half-plane, at $\zeta = \frac{1}{2}\mathrm{i}\lambda \ (\lambda > 0)$, and then $\hat{a}(\zeta)$ has one zero in the lower half-plane at $\zeta = -\frac{1}{2}\mathrm{i}\lambda$. (Similar results hold for $\lambda < 0$, although the corresponding forms of equations (6.30) and (6.31) now differ slightly.) With $\lambda > 0$ then $b(\zeta) = 0$ for all $\zeta \neq \frac{1}{2}i\lambda$, but $b(\frac{1}{2}i\lambda) \neq 0$. This is easily demonstrated by considering equation (6.32) for $\zeta = \frac{1}{2}i\lambda$ with $a = 0$, so that

$$e^{\lambda x/2}\begin{pmatrix} t-1 \\ s \end{pmatrix} = b_1 e^{-\lambda x/2}\begin{pmatrix} s \\ -1-t \end{pmatrix},$$

where $b_1 = b(\zeta_1)$, $\zeta_1 = \frac{1}{2}i\lambda$. This equation holds for all x only if $b_1 = -1$, and so

$$b(\zeta) = 0, \qquad \zeta \neq \tfrac{1}{2}i\lambda; \qquad b(\tfrac{1}{2}i\lambda) = -1.$$

Similarly, we can show that $\hat{b}(-\frac{1}{2}i\lambda) = -1$.

Now let us turn to the inverse problem and, as an exercise, reconstruct the two potentials discussed above. This requires that we start from the appropriate scattering data:

$$a(\zeta) = \frac{2i\zeta + \lambda}{2i\zeta - \lambda}; \qquad\qquad \hat{a}(\zeta) = a^*(\zeta),$$

$$b(\zeta) = 0, \zeta \neq \tfrac{1}{2}i\lambda; \qquad\qquad b(\tfrac{1}{2}i\lambda) = -1,$$

$$\hat{b}(\zeta) = 0, \zeta \neq -\tfrac{1}{2}i\lambda; \qquad b(-\tfrac{1}{2}i\lambda) = -1,$$

for $\lambda > 0$. Thus we can determine the normalisation constants, c_n $(n = 1)$ and $\hat{c}_m$ $(m = 1)$ (see equations (6.26) and (6.27)), as

$$c_1 = (-1)/(2i/\{-2\lambda\}) = -i\lambda; \qquad \hat{c}_1 = (-1)/(2i/2\lambda) = i\lambda,$$

and hence

$$f(X) = \hat{f}(X) = -\lambda e^{-\lambda X/2}.$$

From equation (6.24), because the matrix F is antisymmetric, we obtain the two (scalar) integral equations

$$K_{11}(x, z) - \lambda \int_x^\infty K_{12}(x, y)e^{-\lambda(y+z)/2}\, dy = 0; \qquad (6.35)$$

and

$$\lambda e^{-\lambda(x+z)/2} + K_{12}(x, z) + \lambda \int_x^\infty K_{11}(x, y)e^{-\lambda(y+z)/2}\, dy = 0, \qquad (6.36)$$

with $K_{22} = K_{11}$ and $K_{21} = -K_{12}$; this latter identity implies that $r = -q$ (see equations (6.25)). Equation (6.35) implies that

$$K_{11}(x, x) = L(x)e^{-\lambda z/2}$$

where

$$L(x) = \lambda \int_x^\infty K_{12}(x, y)e^{-\lambda y/2}\, dy. \qquad (6.37)$$

Now from equation (6.36) we obtain

$$K_{12}(x,z) = -\lambda e^{-\lambda(x+z)/2} - \lambda L(x)e^{-\lambda z/2} \int_x^\infty e^{-\lambda y}\,\mathrm{d}y$$

$$= -\{\lambda e^{-\lambda x/2} + L(x)e^{-\lambda x}\}e^{-\lambda z/2}, \qquad (6.38)$$

and so equation (6.37) can be written as

$$L(x) = -\lambda^2 e^{-\lambda x/2}\int_x^\infty e^{-\lambda y}\,\mathrm{d}y - \lambda L(x)e^{-\lambda x}\int_x^\infty e^{-\lambda y}\,\mathrm{d}y$$

$$= -\{\lambda e^{\lambda x/2} + L(x)\}e^{-2\lambda x}.$$

Thus

$$L(x) = -\frac{\lambda e^{-3\lambda x/2}}{1 + e^{-2\lambda x}},$$

and hence equation (6.38) yields

$$K_{12}(x,x) = -\lambda/(e^{\lambda x} + e^{-\lambda x})$$

or

$$q(x) = -2K_{12}(x,x) = \lambda\operatorname{sech}\lambda x.$$

We have therefore reconstructed the two potentials – remembering that $r = -q$ – from the given scattering data.

This concludes the discussion of the direct and inverse scattering problems in a 2×2 system. We now analyse how these ideas are relevant to the solution of certain evolution equations.

6.1.4 Time evolution of the scattering data

So far the potentials, q and r, have been functions only of x. The aim now is to consider the consequences of allowing both q and r to depend upon a parameter t, i.e. to let them *evolve* in time. Thus the eigenvectors, ψ, must now also be functions of t: $\psi = \psi(x;t)$. In the AKNS scheme, the time evolution of ψ is to be given by

$$\psi_t = A\psi$$

(cf. equations (6.3)), where $A(x,t;\zeta)$ is a 2×2 matrix with elements A_{ij} $(i,j = 1,2)$ and for which ζ is *not* a function of t. (This latter assumption is the time invariance of the eigenvalues, already proved in the case of the KdV equation.) Thus $\psi(x;t)$ must satisfy the pair of equations

$$\psi_t = A\psi \qquad \text{and} \qquad \psi_x = R\psi, \qquad (6.39a, b)$$

where $R = \begin{pmatrix} -i\zeta & q \\ r & i\zeta \end{pmatrix}$ and $\mathrm{d}\zeta/\mathrm{d}t = 0$.

It is clear that, given R, A is not an arbitrary matrix. The elements of

A must satisfy certain compatibility conditions in order to ensure that the cross-derivative ψ_{xt} exists. Thus we see that

$$\psi_{xt} = R_t\psi + R\psi_t = (R_t + RA)\psi$$

and

$$\psi_{tx} = A_x\psi + A\psi_x = (A_x + AR)\psi,$$

and so, because $\psi_{xt} = \psi_{tx}$,

$$A_x - R_t + [A, R] = 0,$$

where $[A, R] = AR - RA$ as before. This matrix equation is equivalent to the four equations

$$A_{11x} + rA_{12} - qA_{21} = 0;$$
$$A_{12x} + 2i\zeta A_{12} - q_t + q(A_{11} - A_{22}) = 0;$$
$$A_{21x} - 2i\zeta A_{21} - r_t - r(A_{11} - A_{22}) = 0;$$
$$A_{22x} - (rA_{12} - qA_{21}) = 0.$$

We see immediately from the first and fourth equations that these two are consistent only if $A_{22} = -A_{11}$ (to within a shift by the addition of an arbitrary function of t, which turns out to be irrelevant). Thus we shall work hereafter with the equations

$$A_{11x} + rA_{12} - qA_{21} = 0; \tag{6.40a}$$

$$A_{12x} + 2i\zeta A_{12} - q_t + 2qA_{11} = 0; \tag{6.40b}$$

$$A_{21x} - 2i\zeta A_{21} - r_t - 2rA_{11} = 0. \tag{6.40c}$$

The problem now is one of solving equations (6.40) in order to determine those A_{11}, A_{12} and A_{21} for which equations (6.39) are compatible. It turns out (see §6.1.5) that this calculation itself is only possible if a further compatibility condition is satisfied: this then yields those evolution equations (for q and r) which can be solved by the AKNS scheme. However, it is unnecessary at this stage to proceed further with the analysis of equations (6.40) in order to find the time evolution of the scattering data.

We shall restrict further discussion to the case

$$A_{11} \to \alpha(\zeta), \qquad A_{12} \to 0, \qquad A_{21} \to 0 \qquad \text{as } |x| \to \infty, \tag{6.41}$$

which is consistent with both q and r tending to zero as $|x| \to \infty$. The aim is to determine how a, b, $\hat{a}$ and $\hat{b}$ develop in time, t, if ψ evolves according to equations (6.39), under the constraint of equations (6.40) and (6.41). Now the eigenfunctions introduced by definitions (6.6), and which satisfy the asymptotic conditions (6.4) and (6.5), can no longer be used to describe the time-dependent solutions. (Presumably there will be a time dependence superimposed on the asymptotic behaviours as $x \to \pm \infty$.) The way forward is as follows. The time-evolution equation, (6.39a), becomes

$$\psi_t \sim \begin{pmatrix} \alpha & 0 \\ 0 & -\alpha \end{pmatrix} \psi \qquad \text{as } |x| \to \infty,$$

which has the asymptotic solution

$$\psi(x; t) \sim \begin{pmatrix} e^{\alpha t} & 0 \\ 0 & e^{-\alpha t} \end{pmatrix} f(x) \qquad \text{as } |x| \to \infty,$$

where $f(x)$ is an arbitrary two-vector. We now introduce matrices Φ_+ and Φ_-, defined as for equations (6.6), and which also satisfy conditions (6.4) and (6.5), respectively. Thus we can now express two solutions of the pair of equations (6.39) as

$$\psi = \Phi_+ E \qquad \text{and} \qquad \psi = \Phi_- E \qquad \text{where } E = \begin{pmatrix} e^{\alpha t} & 0 \\ 0 & e^{-\alpha t} \end{pmatrix},$$

with

$$\Phi_- = \Phi_+ S \tag{6.42}$$

(see equation (6.7)).

In order that $\Phi_- E$ be a solution of equation (6.39a), we obtain by direct substitution

$$\Phi_{-t} E + \Phi_- E_t = A\Phi_- E, \tag{6.43}$$

and it is easily shown that $E_t = -\alpha \hat{I} E$ where $\hat{I} = \begin{pmatrix} -1 & 0 \\ 0 & 1 \end{pmatrix}$. On post-multiplication by E^{-1}, equation (6.43) gives

$$\Phi_{-t} = A\Phi_- + \alpha\Phi_- \hat{I}. \tag{6.44}$$

If we now differentiate equation (6.42) with respect to t to give

$$\Phi_{-t} = \Phi_{+t} S + \Phi_+ S_t,$$

and then use equation (6.44), we obtain

$$A\Phi_- + \alpha\Phi_- \hat{I} = \Phi_{+t} S + \Phi_+ S_t.$$

This and equation (6.42) yield

$$A\Phi_+ S + \alpha\Phi_+ S\hat{I} = \Phi_{+t} S + \Phi_+ S_t. \tag{6.45}$$

Finally, we evaluate equation (6.45) as $x \to +\infty$ where

$$\Phi_+ \sim \begin{pmatrix} e^{-i\zeta x} & 0 \\ 0 & e^{i\zeta x} \end{pmatrix} \qquad \text{and} \qquad A \to \begin{pmatrix} \alpha & 0 \\ 0 & -\alpha \end{pmatrix};$$

we also see that $\Phi_{+t} \to 0$ as $x \to +\infty$, and so equation (6.45) gives

$$\begin{pmatrix} \alpha & 0 \\ 0 & -\alpha \end{pmatrix} \begin{pmatrix} a & \hat{b} \\ b & -\hat{a} \end{pmatrix} + \begin{pmatrix} a & \hat{b} \\ b & -\hat{a} \end{pmatrix} \begin{pmatrix} -\alpha & 0 \\ 0 & \alpha \end{pmatrix} = \begin{pmatrix} a_t & \hat{b}_t \\ b_t & -\hat{a}_t \end{pmatrix}.$$

(Note that the asymptotic behaviour of Φ_+ has been cancelled throughout this equation.) Thus we obtain

$$a_t = 0, \qquad \hat{a}_t = 0, \qquad b_t = -2\alpha b, \qquad \hat{b}_t = 2\alpha \hat{b};$$

or

$$a = a(\zeta), \qquad \hat{a} = \hat{a}(\zeta), \qquad (6.46)$$

$$b(\zeta; t) = b_0(\zeta)e^{-2\alpha t} \quad \text{and} \quad \hat{b}(\zeta; t) = \hat{b}_0(\zeta)e^{2\alpha t} \quad \text{for all } t, \quad (6.47)$$

where $b = b_0(\zeta)$, and $\hat{b} = \hat{b}_0(\zeta)$, at $t = 0$.

The time evolution of the scattering data has been determined, and it is as simple as that already derived for the KdV equation (see §4.3). In fact, not only are a and $\hat{a}$ independent of t but this in turn implies that their zeros do not change. Thus we have maintained consistency with the original assumption of $d\zeta/dt = 0$: the discrete eigenvalues are constants of the motion. It can also be shown that an infinity of conserved quantities exists (see Q6.5).

The procedure for solving evolution equations now parallels that for the KdV equation. Given initial profiles for the two potentials (q and r), the scattering data ($a, \hat{a}, b_0, \hat{b}_0$) are determined. Provided $\alpha(\zeta)$ is known (see below), we can therefore obtain $b(\zeta; t)$ and $\hat{b}(\zeta; t)$, and hence reconstruct the potentials at any later time. The only information we lack now is a description of the class of evolution equations (for q and r) which can be solved by the AKNS scheme.

6.1.5 *The evolution equations for q and r*

At this point we return to equations (6.40),

$$\left.\begin{array}{l} A_{11x} + rA_{12} - qA_{21} = 0; \\ A_{12x} + 2i\zeta A_{12} - q_t + 2qA_{11} = 0; \\ A_{21x} - 2i\zeta A_{21} - r_t - 2rA_{11} = 0, \end{array}\right\} \qquad (6.48a, b, c)$$

with the corresponding behaviours (6.41) at infinity

$$A_{11} \to \alpha(\zeta), \qquad A_{12} \to 0, \qquad A_{21} \to 0 \qquad \text{as } |x| \to \infty, \quad (6.49)$$

and examine their solution. Although an operator method is available for the construction of the most general evolution equations (see §6.1.5(c) below), we shall adopt an alternative approach here. Indeed, this approach is more straightforward and is often more useful in deriving specific evolution equations. We proceed by exploiting the fact that equations (6.48) involve the parameter ζ, and are valid for *all* ζ. Thus we could, for example, express the matrix $A(x, t; \zeta)$ as a polynomial in ζ and substitute into equations (6.48). Upon setting the coefficients of the various powers of ζ to zero – remembering that q and r are not functions of ζ – we are able to determine not only A, but also the equations for q and r.

6.1.5(a) Quadratic in ζ

One of the simplest cases to investigate is when A is a quadratic polynomial in ζ; let us write

$$
\left.\begin{array}{ll}
A_{11} = -A_{22} = \alpha_0 + \alpha_1\zeta + \alpha_2\zeta^2; & A_{12} = \beta_0 + \beta_1\zeta + \beta_2\zeta^2; \\
A_{21} = \gamma_0 + \gamma_1\zeta + \gamma_2\zeta^2 &
\end{array}\right\} \quad (6.50)
$$

where α_j, β_j and γ_j $(j = 0, 1, 2)$ are functions of x and t. From equation (6.48a) the coefficients of ζ^n $(n = 0, 1, 2)$ give

$$
\alpha_{jx} + r\beta_j - q\gamma_j = 0, \qquad j = 0, 1, 2, \qquad (6.51\text{a, b, c})
$$

and from equation (6.48b) the coefficients of ζ^n $(n = 0, \ldots, 3)$ yield

$$
\beta_2 = 0, \qquad\qquad\qquad (6.52\text{a})
$$

with

$$
\beta_{jx} + 2\mathrm{i}\beta_{j-1} + 2q\alpha_j = 0 \qquad (j = 1, 2), \qquad (6.52\text{b, c})
$$

and

$$
\beta_{0x} - q_t + 2q\alpha_0 = 0. \qquad\qquad (6.52\text{d})
$$

Similarly, from equation (6.48c), we obtain

$$
\gamma_2 = 0, \qquad\qquad\qquad (6.53\text{a})
$$

$$
\gamma_{jx} - 2\mathrm{i}\gamma_{j-1} - 2r\alpha_j = 0 \qquad (j = 1, 2), \qquad (6.53\text{b, c})
$$

$$
\gamma_{0x} - r_t - 2r\alpha_0 = 0. \qquad\qquad (6.53\text{d})
$$

(We note that this gives eleven equations relating the nine unknowns $\alpha_j, \beta_j, \gamma_j$; thus we may anticipate that q and r are not arbitrary functions.)

Now from equation (6.51c), with the use of equations (6.52a) and (6.53a), we see that $\alpha_2 = \hat\alpha_2(t)$. Here $\hat\alpha_1$ and $\hat\alpha_2$ are arbitrary functions. Also, from equations (6.52c) and (6.53c) we obtain

$$
\beta_1 = \mathrm{i}q\hat\alpha_2 \qquad \text{and} \qquad \gamma_1 = \mathrm{i}r\hat\alpha_2,
$$

respectively, and then equation (6.51b) gives $\alpha_1 = \hat\alpha_1(t)$. Thus we are left with equations (6.51a), (6.52b, d) and (6.53b, d) which become

$$
\alpha_{0x} + r\beta_0 - q\gamma_0 = 0, \qquad\qquad (6.54\text{a})
$$

$$
\mathrm{i}\hat\alpha_2 q_x + 2\mathrm{i}\beta_0 + 2q\hat\alpha_1 = 0, \qquad\qquad (6.54\text{b})
$$

$$
\beta_{0x} - q_t + 2q\alpha_0 = 0, \qquad\qquad (6.54\text{c})
$$

$$
\mathrm{i}\hat\alpha_2 r_x - 2\mathrm{i}\gamma_0 - 2r\hat\alpha_1 = 0, \qquad\qquad (6.54\text{d})
$$

$$
\gamma_{0x} - r_t - 2r\alpha_0 = 0, \qquad\qquad (6.54\text{e})
$$

respectively. Equations (6.54b, d) immediately give

$$\beta_0 = iq\hat{\alpha}_1 - \tfrac{1}{2}\hat{\alpha}_2 q_x \qquad \text{and} \qquad \gamma_0 = ir\hat{\alpha}_1 + \tfrac{1}{2}\hat{\alpha}_2 r_x,$$

respectively, and so equation (6.54a) becomes

$$\alpha_{0x} = \tfrac{1}{2}\hat{\alpha}_2 (qr)_x \qquad \text{or} \qquad \alpha_0 = \tfrac{1}{2}\hat{\alpha}_2 qr + \hat{\alpha}_0,$$

where $\hat{\alpha}_0(t)$ is an arbitrary function. Finally, equations (6.54c, e) can be written as

$$q_t = i\hat{\alpha}_1 q_x - \tfrac{1}{2}\hat{\alpha}_2 q_{xx} + \hat{\alpha}_2 rq^2 + 2\hat{\alpha}_0 q, \tag{6.55a}$$

$$r_t = i\hat{\alpha}_1 r_x + \tfrac{1}{2}\hat{\alpha}_2 r_{xx} - \hat{\alpha}_2 qr^2 + 2\hat{\alpha}_0 r, \tag{6.55b}$$

respectively, which constitute the most general evolution equations for q and r if the matrix A is quadratic in ζ (where $\hat{\alpha}_j(t), j = 0, 1, 2$, may be arbitrarily assigned).

A particularly instructive case is afforded by the choice $\hat{\alpha}_1 = \hat{\alpha}_0 = 0$ and $\hat{\alpha}_2 = 2i$, for then equations (6.55) become

$$iq_t - q_{xx} - 2q|q|^2 = 0, \tag{6.56}$$

with $r = -q^*$; thus q satisfies one form of the NLS equation. The matrix A now has the elements

$$A_{11} = -A_{22} = -i|q|^2 + 2i\zeta^2; \qquad A_{12} = -iq_x - 2q\zeta;$$

$$A_{21} = -iq_x^* + 2q^*\zeta,$$

and so we observe that if $q \to 0$ as $|x| \to \infty$ then

$$A_{11} \to 2i\zeta^2, \qquad A_{12} \to 0, \qquad A_{21} \to 0 \qquad \text{as } |x| \to \infty,$$

and hence $\alpha(\zeta) = 2i\zeta^2$ (see equation (6.41)). (The choice $r = q^*$ gives $iq_t - q_{xx} + 2q|q|^2 = 0$ which, it turns out, does not have any soliton solution for which $|q| \to 0$ as $|x| \to \infty$; see Q6.8.)

6.1.5(b) Polynomial in ζ^{-1}

Another important case arises if we choose the matrix A to be a polynomial in *inverse* powers of ζ (which makes this a *degenerate* problem); in fact the simplest such choice, $A \propto \zeta^{-1}$, proves useful. Let us write

$$A_{11} = -A_{22} = l(x,t)/\zeta; \qquad A_{12} = m(x,t)/\zeta; \qquad A_{21} = n(x,t)/\zeta,$$

and then the coefficients of ζ^0 and ζ^{-1}, from equations (6.48), give

$$l_x + rm - qn = 0;$$

$$2im = q_t; \qquad m_x + 2ql = 0;$$

$$2in = -r_t; \qquad n_x - 2rl = 0.$$

Thus

$$2il_x = -(qr)_t; \qquad q_{xt} = -4ilq; \qquad r_{xt} = -4ilr;$$

and one special case is obtained by setting

$$l = \tfrac{1}{4}\mathrm{i}\cos u, \qquad m = n = \tfrac{1}{4}\mathrm{i}\sin u,$$

provided $q = -r = -\tfrac{1}{2}u_x$ where $u(x,t)$ satisfies

$$u_{xt} = \sin u, \tag{6.57}$$

the sine–Gordon (SG) equation. We can see that

$$A_{11} \to \frac{\mathrm{i}}{4\zeta} \qquad \text{and} \qquad A_{12} \to 0, \qquad A_{21} \to 0,$$

as $u \to 0$, and so $\alpha(\zeta) = \mathrm{i}/(4\zeta)$.

6.1.5(c) *General function of ζ*

In conclusion we now quote the general form of the coupled evolution equations. The details are given in Ablowitz & Segur (1981, §1.5). It can be shown, via an orthogonality condition, that the matrix $A(x,t;\zeta)$ exists only if $q(x,t)$ and $r(x,t)$ satisfy

$$\left\{ -\hat{I}\frac{\partial}{\partial t} + 2\alpha(\mathrm{i}\mathscr{L}) \right\}\binom{r}{q} = 0, \tag{6.58}$$

where $A_{11} \to \alpha(\zeta)$ as $|x| \to \infty$ (see equation (6.41)), and $\mathscr{L}$ is the operator

$$\mathscr{L} = \tfrac{1}{2}\hat{I}\frac{\partial}{\partial x} - \begin{pmatrix} r & 0 \\ 0 & q \end{pmatrix}\int_x^\infty \mathrm{d}x \begin{pmatrix} q & -r \\ q & -r \end{pmatrix}. \tag{6.59}$$

(Note that the integral here operates on *all* functions to the right in equation (6.58).) Furthermore, equation (6.59) gives

$$\mathscr{L} \sim \tfrac{1}{2}\hat{I}\frac{\partial}{\partial x}, \qquad \text{as } x \to +\infty,$$

and so equation (6.58) reads

$$\hat{I}\frac{\partial}{\partial t}\binom{r}{q} \sim 2\alpha\left(\tfrac{1}{2}\mathrm{i}\hat{I}\frac{\partial}{\partial x}\right)\binom{r}{q}, \qquad \text{as } x \to +\infty,$$

which is a pair of *uncoupled linear* equations. Thus for example, if we take the solution of the linear equation for r as

$$r = \mathrm{e}^{\mathrm{i}(kx - \omega t)}, \tag{6.60}$$

where $\omega = \omega(k)$ and $\alpha = \alpha(\zeta)$, then

$$\mathrm{i}\omega(k) = 2\alpha(k/2) \qquad \text{i.e.} \qquad \alpha(\zeta) = \tfrac{1}{2}\mathrm{i}\omega(2\zeta). \tag{6.61}$$

(This follows when we note that with $\alpha(\zeta) = \alpha_0 \zeta^n$, then

$$\alpha\left(\tfrac{1}{2}\mathrm{i}\hat{I}\frac{\partial}{\partial x}\right) = \alpha_0(\tfrac{1}{2}\mathrm{i})^n \hat{I}^n \frac{\partial^n}{\partial x^n}$$

and $\hat{I}^2 = I$, the unit matrix. If both r and q are proportional to e^{ikx}, so that $\begin{pmatrix} r \\ q \end{pmatrix} = e^{ikx} \begin{pmatrix} \hat{r} \\ \hat{q} \end{pmatrix}$, then

$$\alpha\left(\tfrac{1}{2}i\hat{I}\frac{\partial}{\partial x}\right)\begin{pmatrix} r \\ q \end{pmatrix} = \alpha\left(\tfrac{1}{2}i\hat{I}\frac{\partial}{\partial x}\right)e^{ikx}\begin{pmatrix} \hat{r} \\ \hat{q} \end{pmatrix} = \alpha_0(-\tfrac{1}{2})^n k^n \hat{I}^n \begin{pmatrix} \hat{r} \\ \hat{q} \end{pmatrix},$$

and equations (6.61) follow for the component r.)

Thus $\alpha(\zeta)$ is related very simply to the dispersion function, $\omega(k)$, of the underlying linear equation; a corresponding result is obtained for q. Of course, it is the choice of $\alpha(\zeta)$ which now determines the evolution equations for q and r from equation (6.58). We must note, however, that this choice extends only to the linear problem and that the nonlinear system described by equation (6.58) is then completely prescribed. For example, if we take the linear equation

$$ir_t + r_{xx} = 0$$

for which $\omega(k) = k^2$, equation (6.58) then becomes

$$\left\{-\hat{I}\frac{\partial}{\partial t} + i(2i\mathscr{L})^2\right\}\begin{pmatrix} r \\ q \end{pmatrix} = 0.$$

This can be simplified, after a little calculation, to give

$$\left\{\hat{I}\frac{\partial}{\partial t} + \begin{pmatrix} \partial^2/\partial x^2 & -2r^2 \\ -2q^2 & \partial^2/\partial x^2 \end{pmatrix}\right\}\begin{pmatrix} r \\ q \end{pmatrix} = 0,$$

which with $r = -q^*$ is the NLS equation, (6.56).

6.2 The ZS scheme

6.2.1 The integral operators

Zakharov & Shabat (1974), as we have already mentioned, generalised the Lax method. We shall see the connection with the work of Lax soon, but first we introduce three integral operators. Let $F(x, z)$ and $K_{\pm}(x, z)$ be $N \times N$ matrices where

$$K_+(x, z) = 0 \qquad \text{if } z < x, \tag{6.62a}$$

and

$$K_-(x, z) = 0 \qquad \text{if } z > x, \tag{6.62b}$$

and let $\psi(x)$ be an N-vector. (We have chosen a notation similar to that used earlier, although a direct correspondence is not implied at this stage.) The integral operators J_F and $J_{\pm}$ on ψ are defined by

$$J_F(\psi) = \int_{-\infty}^{\infty} F(x, z)\psi(z) \, dz \tag{6.63}$$

for all integrable ψ, and similarly

$$\mathbf{J}_\pm(\psi) = \int_{-\infty}^\infty K_\pm(x,z)\psi(z)\,dz, \tag{6.64}$$

so that

$$\mathbf{J}_+ = \int_x^\infty dz\,K_+(x,z) \quad\text{and}\quad \mathbf{J}_- = \int_{-\infty}^x dz\,K_-(x,z).$$

We now suppose that $\mathbf{J}_F$ and $\mathbf{J}_\pm$ are related by the operator identity

$$(I + \mathbf{J}_+)(I + \mathbf{J}_F) = I + \mathbf{J}_-, \tag{6.65}$$

where we assume that $I + \mathbf{J}_+$ is *invertible* so that

$$I + \mathbf{J}_F = (I + \mathbf{J}_+)^{-1}(I + \mathbf{J}_-),$$

i.e. the operator $I + \mathbf{J}_F$ is factorisable. (I, as usual, is the unit matrix.)
The identity (6.65), on ψ, can be written as

$$\int_x^\infty K_+(x,z)\psi(z)\,dz + \int_{-\infty}^\infty F(x,z)\psi(z)\,dz$$

$$+ \int_x^\infty K_+(x,z)\left(\int_{-\infty}^\infty F(z,y)\psi(y)\,dy\right)dz = \int_{-\infty}^x K_-(x,z)\psi(z)\,dz, \tag{6.66}$$

and if we choose to operate in $z > x$ then the right-hand side is zero.
Furthermore, the double integral may be expressed as

$$\int_{-\infty}^\infty \int_x^\infty K_+(x,y)F(y,z)\psi(z)\,dy\,dz,$$

where the 'dummy' variables have been relabelled by interchanging y and
z. Thus equation (6.66) becomes

$$\int_{-\infty}^\infty \left\{ K_+(x,z) + F(x,z) + \int_x^\infty K_+(x,y)F(y,z)\,dy \right\}\psi(z)\,dz = 0,$$

(note the use here of equation (6.62a)) for all $\psi(z)$. Therefore

$$K_+(x,z) + F(x,z) + \int_x^\infty K_+(x,y)F(y,z)\,dy = 0, \tag{6.67}$$

for $z > x$. Equation (6.67) is the matrix Marchenko equation for $K_+(x,z)$;
cf. equation (6.24). Similarly, if we consider $z < x$ in equation (6.66), it can
be shown (see Q6.10) that

$$K_-(x,z) = F(x,z) + \int_x^\infty K_+(x,y)F(y,z)\,dy, \tag{6.68}$$

which defines K_- in terms of K_+ and F. At this stage we have not restricted
the choice of F, and this we will next do.

6.2.2 *The differential operators*

In common with our previous analyses, we extend the definitions of the matrices F and $K_{\pm}$, and the vector ψ, so that they all may now depend upon auxiliary variables, e.g. t, y. We shall describe the evolution of F and $K_{\pm}$ (in t, y), and hence relate them to certain evolution equations by introducing appropriate (linear) differential operators. We define the $N \times N$ matrix differential operator Δ_0 on $\psi(x; t, y)$ which has only *constant* coefficients and which commutes with the integral operator J_F, i.e.

$$[\Delta_0, J_F] = \Delta_0 J_F - J_F \Delta_0 = 0. \tag{6.69}$$

(Note that in the term $J_F \Delta_0$, Δ_0 operates on $\psi(x; t, y)$ first and then this is evaluated on $x = z$ for the application of the operator J_F.) Further, we introduce an associated differential operator, Δ, which is defined by the operator identity

$$\Delta(I + J_+) = (I + J_+)\Delta_0. \tag{6.70}$$

(It can be shown that equation (6.70) also holds if J_+ is replaced by J_-; see Q6.11.) The operator Δ_0 is sometimes referred to as 'undressed', and Δ as the 'dressed' operator.

Before the general development, we shall consider an example which will illuminate the meaning of equations (6.69) and (6.70). Let

$$\Delta_0 = I\left(\alpha \frac{\partial}{\partial t} - \frac{\partial^2}{\partial x^2}\right)$$

where α is a constant scalar, and I is the $N \times N$ unit matrix. Thus equation (6.69), when operated on $\psi(x; t)$, becomes

$$\left(\alpha \frac{\partial}{\partial t} - \frac{\partial^2}{\partial x^2}\right)\int_{-\infty}^{\infty} F(x, z; t)\psi(z; t)\,\mathrm{d}z$$

$$-\int_{-\infty}^{\infty} F(x, z; t)\left(\alpha \frac{\partial}{\partial t} - \frac{\partial^2}{\partial z^2}\right)\psi(z; t)\,\mathrm{d}z = 0. \tag{6.71}$$

After integration by parts, it follows that

$$\int_{-\infty}^{\infty} F\psi_{zz}\,\mathrm{d}z = \int_{-\infty}^{\infty} F_{zz}\psi\,\mathrm{d}z,$$

for all bounded continuously twice differentiable ψ provided $\psi, \psi_z \to 0$ as $|z| \to \infty$. Hence equation (6.71) can be written as

$$\int_{-\infty}^{\infty} (\alpha F_t - F_{xx} + F_{zz})\psi\,\mathrm{d}z = 0,$$

and therefore equation (6.69) is an operator identity only if $F(x, z; t)$ satisfies

$$\alpha F_t + F_{zz} - F_{xx} = 0. \tag{6.72}$$

The associated operator Δ is now obtained from

$$\Delta\left\{\psi(x;t) + \int_x^\infty K_+(x,z;t)\psi(z;t)\,dz\right\}$$

$$= \alpha\psi_t - \psi_{xx} + \int_x^\infty K_+(x,z;t)\left(\alpha\frac{\partial}{\partial t} - \frac{\partial^2}{\partial z^2}\right)\psi(z;t)\,dz. \qquad (6.73)$$

Again, we integrate by parts to find

$$\int_x^\infty K_+\psi_{zz}\,dz = -\hat{K}_+\psi_x + \hat{K}_{+z}\psi + \int_x^\infty K_{+zz}\psi\,dz,$$

where $\hat{K}_+ = K_+(x,x;t)$, and we assume that $K_+, K_{+z} \to 0$ as $z \to +\infty$. It is now convenient to set $\Delta = \Delta_0 + \Delta_1$, so that equation (6.73) becomes

$$\Delta_1\left(\psi + \int_x^\infty K_+\psi\,dz\right) + \alpha\int_x^\infty (K_{+t}\psi + K_+\psi_t)\,dz + \psi\frac{d}{dx}\hat{K}_+$$

$$+ \hat{K}_+\psi_x + \hat{K}_{+x}\psi - \int_x^\infty K_{+xx}\psi\,dz$$

$$= \alpha\int_x^\infty K_+\psi_t\,dz + \hat{K}_+\psi_x - \hat{K}_{+z}\psi - \int_x^\infty K_{+zz}\psi\,dz$$

or

$$\left(\Delta_1 + 2\frac{d}{dx}\hat{K}_+\right)\psi + \Delta_1\int_x^\infty K_+\psi\,dz + \int_x^\infty (\alpha K_{+t} - K_{+xx} + K_{+zz})\psi\,dz = 0$$

since $d\hat{K}_+/dx = \hat{K}_{+x} + \hat{K}_{+z}$, the total derivative in x. Hence if this equation is valid for all continuous ψ, then

$$\Delta_1(x;t) = -2\frac{d}{dx}\hat{K}_+ = -2(\hat{K}_{+x} + \hat{K}_{+z}) \qquad (6.74)$$

(so that Δ_1 is of degree zero), and $K_+(x,z;t)$ satisfies

$$\alpha K_{+t} + K_{+zz} - K_{+xx} + \Delta_1 K_+ = 0. \qquad (6.75)$$

(Equations (6.72), (6.74) and (6.75) should be compared with those discussed in Q4.1.)

We now return to the main development and describe a further important step in the ZS scheme. This is to introduce *two* pairs of operators Δ_0 and Δ. A typical choice is

$$\Delta_0^{(1)} = I\alpha\frac{\partial}{\partial t} - M_0; \qquad \Delta_0^{(2)} = I\beta\frac{\partial}{\partial y} + L_0,$$

and $\qquad\qquad\qquad\qquad\qquad\qquad\qquad\qquad\qquad\qquad\qquad\qquad (6.76)$

$$\Delta^{(1)} = I\alpha\frac{\partial}{\partial t} - M; \qquad \Delta^{(2)} = I\beta\frac{\partial}{\partial y} + L,$$

where α and β are constants, and M_0, L_0, M and L are differential operators in x only. Consistent with our notation, M_0 and L_0 are comprised of constant coefficients only and so $\Delta_0^{(1)}$, $\Delta_0^{(2)}$ commute. Furthermore, both $\Delta_0^{(1)}$ and $\Delta_0^{(2)}$ are to commute with the *same* operator J_F so that

$$[\Delta_0^{(1)}, J_F] = 0 \qquad \text{and} \qquad [\Delta_0^{(2)}, J_F] = 0. \qquad (6.77)$$

The operators $\Delta^{(i)}$ are defined according to equation (6.70), with the *same* J_+,

$$\Delta^{(i)}(I + J_+) = (I + J_+)\Delta_0^{(i)}, \qquad i = 1, 2. \qquad (6.78)$$

At this point it is instructive to examine the operator

$$P = \Delta^{(1)}\Delta^{(2)}(I + J_+) - \Delta^{(2)}\Delta^{(1)}(I + J_+) \qquad (6.79)$$

which, upon the use of equation (6.78) twice, gives

$$\begin{aligned} P &= \Delta^{(1)}(I + J_+)\Delta_0^{(2)} - \Delta^{(2)}(I + J_+)\Delta_0^{(1)} \\ &= (I + J_+)\Delta_0^{(1)}\Delta_0^{(2)} - (I + J_+)\Delta_0^{(2)}\Delta_0^{(1)} \\ &= (I + J_+)[\Delta_0^{(1)}, \Delta_0^{(2)}]. \end{aligned}$$

However, $\Delta_0^{(1)}$ and $\Delta_0^{(2)}$ are chosen so that they commute with one another (see Q6.12); hence $P = 0$. Thus we obtain

$$P = [\Delta^{(1)}, \Delta^{(2)}](I + J_+) = 0$$

and, since $I + J_+$ is invertible (an earlier assumption which can be checked in specific cases), we have

$$[\Delta^{(1)}, \Delta^{(2)}] = 0, \qquad (6.80)$$

i.e. $\Delta^{(1)}$ commutes with $\Delta^{(2)}$. If we now introduce the choice given in equations (6.76), equation (6.80) becomes

$$\left(I\alpha\frac{\partial}{\partial t} - M \right)\left(I\beta\frac{\partial}{\partial y} + L \right) - \left(I\beta\frac{\partial}{\partial y} + L \right)\left(I\alpha\frac{\partial}{\partial t} - M \right) = 0,$$

which simplifies to

$$\alpha\frac{\partial L}{\partial t} + \beta\frac{\partial M}{\partial y} + [L, M] = 0. \qquad (6.81)$$

This is a generalisation of Lax's equation (5.28) to two auxiliary variables; the Lax equation is recovered if $\beta = 0$ and $\alpha = 1$. Equation (6.81) is the system of nonlinear evolution equations which can be solved by the ZS scheme.

The procedure for solving equation (6.81) is now described. The variable coefficients which arise in the 'dressed' operators, L and M, constitute the functions which satisfy the system of evolution equations. These functions are known in terms of K_+ (cf. equation (6.74)), where K_+ is a solution of the linear integral equation, (6.67). This equation requires F, and F is supplied by the solution of a pair of equations (cf. equation (6.72)), a *pair* since both equations (6.77) are to be satisfied. Note that the eigenvalue does not appear explicitly at any stage in the ZS scheme.

The correspondence between the AKNS scheme and the ZS scheme is now clear. In the first place they both incorporate the same form of matrix Marchenko equation (cf. equations (6.24) and (6.67)), although the AKNS scheme is a 2×2 system whereas the ZS is $N \times N$. However, the definitions of the two F functions differ significantly (see equations (6.22) and (6.77)). In fact, the definition of F in the ZS scheme, via the linear partial differential equations implied by equations (6.77), is often more useful in practice. In the second place the time evolution of F is particularly simple in both schemes (from equations (6.46) and (6.47), and one of the pair (6.77)), but the representations of the evolution equations which can be solved do differ somewhat (cf. equations (6.58) and (6.81)).

To conclude this discussion of the ZS scheme we shall present some examples which show how various standard evolution equations can be obtained. In §6.3 we shall then describe two simple solutions obtained by employing the schemes described here.

6.2.3 Scalar operators

The simplest choice for the pair of operators $\Delta_0^{(1)}$ and $\Delta_0^{(2)}$ is to let them be scalars. With this restriction we shall examine two examples, the first of which recovers the classical KdV equation.

6.2.3(a) The KdV equation

In this case we set

$$\Delta_0^{(1)} = \frac{\partial}{\partial t} + 4\frac{\partial^3}{\partial x^3} \qquad \text{and} \qquad \Delta_0^{(2)} = -\frac{\partial^2}{\partial x^2}$$

(see equations (6.76) with $\alpha = 1$, $M_0 = -4\partial^3/\partial x^3$, $\beta = 0$ and $L_0 = -\partial^2/\partial x^2$), and write

$$\Delta^{(2)} = L = -\frac{\partial^2}{\partial x^2} + u(x, t)$$

so that from equations (6.74) and (6.75) we obtain

$$u(x, t) = -2\frac{\mathrm{d}}{\mathrm{d}x}K_+(x, x; t) \qquad \text{where } K_{+zz} - K_{+xx} + uK_+ = 0, \quad (6.82)$$

(since Δ_1 is replaced here by u). Similarly, we see that $[\Delta_0^{(2)}, J_F] = 0$ implies

$$F_{xx} - F_{zz} = 0, \qquad (6.83)$$

since $\alpha = 0$ in equation (6.72). The identity $[\Delta_0^{(1)}, J_F] = 0$ gives

$$F_t + 4(F_{xxx} + F_{zzz}) = 0, \qquad (6.84)$$

obtained by following the analysis which leads to equations (6.72); this is left as an exercise (see Q6.13). Finally, we have

$$\Delta^{(1)} = \frac{\partial}{\partial t} - M$$

where we write $M = M_0 + M_1$, and $M_0 = -4\partial^3/\partial x^3$, so that equation (6.78) on ψ becomes

$$-M_1\psi + \left(\frac{\partial}{\partial t} + 4\frac{\partial^3}{\partial x^3} - M_1\right)\int_x^\infty K_+\psi\,dz = \int_x^\infty K_+\left(\frac{\partial}{\partial t} + 4\frac{\partial^3}{\partial z^3}\right)\psi\,dz.$$

This simplifies to give

$$-M_1\psi + \int_x^\infty K_{+t}\psi\,dz + 4\frac{\partial^3}{\partial x^3}\int_x^\infty K_+\psi\,dz - M_1\int_x^\infty K_+\psi\,dz$$

$$= 4\int_x^\infty K_+\psi_{zzz}\,dz,$$

which in turn becomes

$$-M_1\left(\psi + \int_x^\infty K_+\psi\,dz\right) + \int_x^\infty K_{+t}\psi\,dz - 4\left(\frac{d^2\hat{K}_+}{dx^2} + \frac{d\hat{K}_{+x}}{dx} + \hat{K}_{+xx}\right)\psi$$

$$-4\left(2\frac{d\hat{K}_+}{dx} + \hat{K}_{+x}\right)\psi_x - 4\hat{K}_+\psi_{xx} + 4\int_x^\infty K_{+xxx}\psi\,dz$$

$$= -4\hat{K}_+\psi_{xx} + 4\hat{K}_{+z}\psi_x - 4\hat{K}_{+zz}\psi - 4\int_x^\infty K_{+zzz}\,\psi\,dz, \qquad (6.85)$$

after expanding the term in $\partial^3/\partial x^3$ and integrating by parts on the term in ψ_{zzz}. (Remember that $\hat{K}_+ = K_+(x, x; t)$.) Now suppose that

$$M_1 = 4A(x, t)\frac{\partial}{\partial x} + 4B(x, t), \qquad (6.86)$$

then equation (6.85) can be written as

$$-4\left(\frac{d^2\hat{K}_+}{dx^2} + \frac{d\hat{K}_{+x}}{dx} + \hat{K}_{+xx} - \hat{K}_{+zz} + B - A\hat{K}_+\right)\psi - 4\left(3\frac{d\hat{K}_+}{dx} + A\right)\psi_x$$

$$+ \int_x^\infty (K_{+t} + 4K_{+xxx} + 4K_{+zzz} - 4BK_+ - 4AK_{+x})\psi\,dz = 0.$$

Thus we choose

$$A = -3\frac{d\hat{K}_+}{dx} = \tfrac{3}{2}u, \qquad (6.87)$$

and

$$B = A\hat{K}_+ + \hat{K}_{+zz} - \hat{K}_{+xx} - \frac{d\hat{K}_{+x}}{dx} - \frac{d^2\hat{K}_+}{dx^2}$$

$$= \tfrac{3}{2}(\hat{K}_{+xx} - \hat{K}_{+zz}) + \hat{K}_{+zz} - \hat{K}_{+xx} - (\hat{K}_{+xx} + \hat{K}_{+xz}) - \frac{d^2\hat{K}_+}{dx^2}$$

if we use equations (6.87) and (6.82). Hence

$$B = -\frac{3}{2}\frac{d^2\hat{K}_+}{dx^2} = \tfrac{3}{4}u_x$$

and so

$$\Delta^{(1)} = \frac{\partial}{\partial t} + 4\frac{\partial^3}{\partial x^3} - 6u\frac{\partial}{\partial x} - 3u_x.$$

The resulting evolution equation obtained from equation (6.81), with

$$\alpha = 1, \qquad \beta = 0, \qquad L = -\frac{\partial^2}{\partial x^2} + u$$

and

$$M = -4\frac{\partial^3}{\partial x^3} + 6u\frac{\partial}{\partial x} + 3u_x,$$

is

$$u_t + \left(-\frac{\partial^2}{\partial x^2} + u \right)\left(-4\frac{\partial^3}{\partial x^3} + 6u\frac{\partial}{\partial x} + 3u_x \right)$$

$$- \left(-4\frac{\partial^3}{\partial x^3} + 6u\frac{\partial}{\partial x} + 3u_x \right)\left(-\frac{\partial^2}{\partial x^2} + u \right) = 0,$$

or

$$u_t - 6uu_x + u_{xxx} = 0,$$

our usual form of the KdV equation.

6.2.3(b) The two-dimensional KdV equation

This example is a straightforward extension of the choice required for the KdV equation; it involves the addition of a second auxiliary variable, y. Thus we choose

$$\Delta_0^{(1)} = \frac{\partial}{\partial t} + 4\frac{\partial^3}{\partial x^3} \qquad \text{and} \qquad \Delta_0^{(2)} = \frac{\partial}{\partial y} - \frac{\partial^2}{\partial x^2},$$

with

$$\Delta^{(2)} = \Delta_0^{(2)} + u(x, t, y)$$

and

$$\Delta^{(1)} = \Delta_0^{(1)} - 6u\frac{\partial}{\partial x} - 3u_x + w(x, t, y).$$

If we follow the development given in §6.2.3(a) we find that

$$F_t + 4(F_{xxx} + F_{zzz}) = 0$$

and

$$F_y + F_{zz} - F_{xx} = 0,$$

with

$$u_t - 6uu_x + u_{xxx} - w_y = 0,$$

where

$$w_x = -3u_y.$$

The latter two equations can be written, upon the elimination of w, as

$$(u_t - 6uu_x + u_{xxx})_x + 3u_{yy} = 0,$$

namely the 2D KdV (or Kadomtsev–Petviashvili) equation (see Q5.14, Q6.14).

6.2.4 Matrix operators

Since the operators $\Delta_0^{(1)}$ and $\Delta_0^{(2)}$ may be matrices, we anticipate that a wide class of evolution equations can be found. In this section we shall give two examples which show how the ZS scheme works for matrix operators.

6.2.4(a) The nonlinear Schrödinger equation

We start by considering

$$\Delta_0^{(1)} = I\left(i\alpha \frac{\partial}{\partial t} - \frac{\partial^2}{\partial x^2} \right) \qquad \text{and} \qquad \Delta_0^{(2)} = \begin{pmatrix} l & 0 \\ 0 & m \end{pmatrix} \frac{\partial}{\partial x},$$

where l, m and α are real constants, and I is the 2×2 unit matrix. From equation (6.74) we see that

$$\Delta^{(1)} = I\left(i\alpha \frac{\partial}{\partial t} - \frac{\partial^2}{\partial x^2} \right) + U(x, t)$$

where $U = -2\,d\hat{K}_+/dx$, and from (6.78) we obtain

$$\begin{pmatrix} l & 0 \\ 0 & m \end{pmatrix}\left(\psi_x - \hat{K}_+\psi + \int_x^\infty K_{+x}\psi\,dz \right) + V\left(\psi + \int_x^\infty K\psi\,dz \right)$$
$$= \begin{pmatrix} l & 0 \\ 0 & m \end{pmatrix}\psi_x + \int_x^\infty K_+\begin{pmatrix} l & 0 \\ 0 & m \end{pmatrix}\psi_z\,dz, \qquad (6.88)$$

where we have written $\Delta^{(2)} = \Delta_0^{(2)} + V(x, t)$. On integrating by parts in the last term, equation (6.88) becomes

$$\left\{V - \begin{pmatrix} l & 0 \\ 0 & m \end{pmatrix} \hat{K}_+ + \hat{K}_+ \begin{pmatrix} l & 0 \\ 0 & m \end{pmatrix}\right\} \psi$$

$$+ \int_x^\infty \left\{\begin{pmatrix} l & 0 \\ 0 & m \end{pmatrix} K_{+x} + K_{+z} \begin{pmatrix} l & 0 \\ 0 & m \end{pmatrix}\right\} \psi \, dz + V \int_x^\infty K_+ \psi \, dz = 0,$$

and so we choose

$$V = \begin{pmatrix} l & 0 \\ 0 & m \end{pmatrix} \hat{K}_+ - \hat{K}_+ \begin{pmatrix} l & 0 \\ 0 & m \end{pmatrix} = (l - m) \begin{pmatrix} 0 & B \\ -C & 0 \end{pmatrix}$$

where $\hat{K}_+ = \begin{pmatrix} A & B \\ C & D \end{pmatrix}$. Thus $K_+(x, z; t)$ must satisfy

$$\begin{pmatrix} l & 0 \\ 0 & m \end{pmatrix} K_{+x} + K_{+z} \begin{pmatrix} l & 0 \\ 0 & m \end{pmatrix} + VK_+ = 0,$$

and if this equation is evaluated on $z = x$ we find, for example, that

$$lA_x = -(l - m)BC \qquad \text{and} \qquad mD_x = (l - m)BC.$$

The equations for $F(x, z; t)$, from equations (6.77), are

$$\begin{pmatrix} l & 0 \\ 0 & m \end{pmatrix} F_x + F_z \begin{pmatrix} l & 0 \\ 0 & m \end{pmatrix} = 0$$

and

$$i\alpha F_t + F_{zz} - F_{xx} = 0.$$

An equation for $u(x, t)$ can be obtained if we let $B = u$ and $C = \pm u^*$ (u^* is the complex conjugate), so that $BC = \pm |u|^2$ which gives

$$\Delta^{(1)} = I\left(i\alpha \frac{\partial}{\partial t} - \frac{\partial^2}{\partial x^2}\right) - 2\begin{pmatrix} \mp(l-m)|u|^2/l & u_x \\ \pm u_x^* & \pm(l-m)|u|^2/m \end{pmatrix},$$

and

$$\Delta^{(2)} = \begin{pmatrix} l & 0 \\ 0 & m \end{pmatrix} \frac{\partial}{\partial x} + (l - m)\begin{pmatrix} 0 & u \\ \mp u^* & 0 \end{pmatrix}.$$

The Lax equation, (6.81), with $\alpha \to i\alpha$ and $\beta = 0$, therefore becomes

$$i\alpha(l - m)\begin{pmatrix} 0 & u_t \\ \mp u_t^* & 0 \end{pmatrix} + \left\{\begin{pmatrix} l & 0 \\ 0 & m \end{pmatrix} \frac{\partial}{\partial x} + (l - m)\begin{pmatrix} 0 & u \\ \mp u^* & 0 \end{pmatrix}\right\}$$

$$\times \left\{I \frac{\partial^2}{\partial x^2} + 2\begin{pmatrix} \mp(l-m)|u|^2/l & u_x \\ \pm u_x^* & \pm(l-m)|u|^2/m \end{pmatrix}\right\}$$

$$- \left\{I \frac{\partial^2}{\partial x^2} + 2\begin{pmatrix} \mp(l-m)|u|^2/l & u_x \\ \pm u_x^* & \pm(l-m)|u|^2/m \end{pmatrix}\right\}$$

$$\times \left\{\begin{pmatrix} l & 0 \\ 0 & m \end{pmatrix} \frac{\partial}{\partial x} + (l - m)\begin{pmatrix} 0 & u \\ \mp u^* & 0 \end{pmatrix}\right\} = 0 \qquad (6.89)$$

which, after expanding the operators, implies that

$$i\alpha(l-m)u_t + (l+m)u_{xx} \pm \frac{2}{lm}(l-m)(l^2-m^2)u|u|^2 = 0. \qquad (6.90)$$

This is a nonlinear Schrödinger equation, with coefficients which depend on the arbitrary constants l, m and α. (Two elements in the matrix equation (6.89) are identically zero, one gives equation (6.90) and the fourth is the complex conjugate of equation (6.90).)

6.2.4(b) The sine–Gordon equation

This example exploits the matrix formulation more fully by using 4×4 matrices. Let us define

$$\Delta_0^{(1)} = I\alpha\frac{\partial}{\partial t} + \begin{pmatrix} -I & 0 \\ 0 & I \end{pmatrix}\frac{\partial}{\partial x} \qquad \text{and} \qquad \Delta_0^{(2)} = \begin{pmatrix} \tilde{I} & 0 \\ 0 & 0 \end{pmatrix}\frac{\partial}{\partial x}, \qquad (6.91)$$

where I is the unit matrix (either 4×4 or 2×2), $\tilde{I} = \begin{pmatrix} 0 & -1 \\ 1 & 0 \end{pmatrix}$ and α is a real constant. (The zeros appearing in the definitions (6.91) represent 2×2 zero matrices, so that we have partitioned the operators into four 2×2 matrices.) We may now obtain $\Delta^{(1)}$ and $\Delta^{(2)}$ from a single calculation: let

$$\Delta_0 = I\alpha\frac{\partial}{\partial t} + I_0\frac{\partial}{\partial x}$$

so that I_0 can be chosen as the matrix used in $\Delta_0^{(1)}$ or, with $\alpha = 0$, as that in $\Delta_0^{(2)}$. Now we write

$$\Delta = \Delta_0 + W(x,t)$$

so that equation (6.78), operating on ψ, becomes

$$\Delta_0\left(\psi + \int_x^\infty K_+\psi\,dz\right) + W\left(\psi + \int_x^\infty K_+\psi\,dz\right) = \Delta_0\psi + \int_x^\infty K_+(\Delta_0\psi)\,dz$$

or

$$\alpha\int_x^\infty K_{+t}\psi\,dz + I_0\left(-\hat{K}_+\psi + \int_x^\infty K_{+x}\psi\,dz\right) + W\left(\psi + \int_x^\infty K_+\psi\,dz\right)$$

$$= \int_x^\infty K_+I_0\psi_z\,dz = -\hat{K}_+I_0\psi - \int_x^\infty K_{+z}I_0\psi\,dz,$$

provided that $K_+(x,z;t) \to 0$ as $z \to +\infty$. (Remember that $\hat{K}_+ = \hat{K}_+(x;t) = K_+(x,x;t)$.) This equation is satisfied if

$$W + [\hat{K}_+, I_0] = 0, \qquad (6.92)$$

where $K_+(x, z; t)$ is a solution of

$$\alpha K_{+t} + I_0 K_{+x} + K_{+z} I_0 + W K_+ = 0. \tag{6.93}$$

Thus if

$$\Delta^{(1)} = \Delta_0^{(1)} + W_1(x, t)$$

then equation (6.92) yields

$$W_1 = \begin{pmatrix} -I & 0 \\ 0 & I \end{pmatrix} \begin{pmatrix} \hat{A} & \hat{B} \\ \hat{C} & \hat{D} \end{pmatrix} - \begin{pmatrix} \hat{A} & \hat{B} \\ \hat{C} & \hat{D} \end{pmatrix} \begin{pmatrix} -I & 0 \\ 0 & I \end{pmatrix}$$

$$= \begin{pmatrix} 0 & -2\hat{B} \\ 2\hat{C} & 0 \end{pmatrix},$$

where $K_+ = \begin{pmatrix} A & B \\ C & D \end{pmatrix}$. Similarly, if

$$\Delta^{(2)} = \Delta_0^{(2)} + W_2(x, t)$$

then

$$W_2 = \begin{pmatrix} \tilde{I} & 0 \\ 0 & 0 \end{pmatrix} \begin{pmatrix} \hat{A} & \hat{B} \\ \hat{C} & \hat{D} \end{pmatrix} - \begin{pmatrix} \hat{A} & \hat{B} \\ \hat{C} & \hat{D} \end{pmatrix} \begin{pmatrix} \tilde{I} & 0 \\ 0 & 0 \end{pmatrix}$$

$$= \begin{pmatrix} [\tilde{I}, \hat{A}] & \tilde{I}\hat{B} \\ -\hat{C}\tilde{I} & 0 \end{pmatrix}.$$

The evolution equation (6.81), with $\beta = 0$, is therefore

$$\alpha W_{2t} + \left[\left\{ \begin{pmatrix} -I & 0 \\ 0 & I \end{pmatrix} \frac{\partial}{\partial x} + W_1 \right\}, \left\{ \begin{pmatrix} \tilde{I} & 0 \\ 0 & 0 \end{pmatrix} \frac{\partial}{\partial x} + W_2 \right\} \right] = 0$$

which can be simplified to read

$$\alpha \begin{pmatrix} [\tilde{I}, \hat{A}] & \tilde{I}\hat{B} \\ -\hat{C}\tilde{I} & 0 \end{pmatrix}_t + \begin{pmatrix} -[\tilde{I}, \hat{A}] & \tilde{I}\hat{B} \\ -\hat{C}\tilde{I} & 0 \end{pmatrix}_x + 2 \begin{pmatrix} [\hat{B}\hat{C}, \tilde{I}] & [\tilde{I}, \hat{A}]\hat{B} \\ \hat{C}[\tilde{I}, \hat{A}] & 0 \end{pmatrix} = 0. \tag{6.94}$$

We now choose

$$[\tilde{I}, \hat{A}] = \frac{i}{4} \begin{pmatrix} 0 & w \\ w & 0 \end{pmatrix}, \qquad w = w(x, t),$$

and

$$\tilde{I}\hat{B} = -\hat{C}\tilde{I} = v = \frac{1}{4} \begin{pmatrix} e^{iu/2} & 0 \\ 0 & e^{-iu/2} \end{pmatrix}, \qquad u = u(x, t).$$

(These require $\hat{B} = -\tilde{I}v$ and $\hat{C} = v\tilde{I}$, since $\tilde{I}^2 = -I$, and $\hat{A} = \begin{pmatrix} a & b \\ -b & a - iw/4 \end{pmatrix}$ for arbitrary a and b.) Equation (6.94) therefore yields

$$\frac{i}{4} \left(\alpha \frac{\partial}{\partial t} - \frac{\partial}{\partial x} \right) \begin{pmatrix} 0 & w \\ w & 0 \end{pmatrix} + 2[\tilde{I}, v^2] = 0 \tag{6.95}$$

and

$$(\alpha v_t + v_x)I - \frac{i}{8}\begin{pmatrix} we^{iu/2} & 0 \\ 0 & -we^{-iu/2} \end{pmatrix} = 0, \qquad (6.96)$$

since the equations for $\tilde{I}\hat{B}$ and $-\hat{C}\tilde{I}$ are identical: both give rise to equation (6.96). Equation (6.95) now becomes

$$\left(\alpha\frac{\partial}{\partial t} - \frac{\partial}{\partial x}\right)\begin{pmatrix} 0 & w \\ w & 0 \end{pmatrix} + (\sin u)\begin{pmatrix} 0 & 1 \\ 1 & 0 \end{pmatrix} = 0$$

which implies that

$$\alpha w_t - w_x + \sin u = 0,$$

and equation (6.96) gives

$$\alpha u_t + u_x = w.$$

Thus, upon the elimination of w, u is a solution of

$$u_{xx} - \alpha^2 u_{tt} = \sin u; \qquad (6.97)$$

this is the sine–Gordon (SG) equation written in laboratory coordinates. Alternatively, if we introduce the characteristic coordinates

$$\xi = \tfrac{1}{2}(x - t/\alpha), \qquad \eta = \tfrac{1}{2}(x + t/\alpha) \qquad (6.98)$$

then equation (6.97) becomes

$$u_{\xi\eta} = \sin u; \qquad (6.99)$$

cf. equation (5.67).

The equations for $K_+ = \begin{pmatrix} A & B \\ C & D \end{pmatrix}$, derived from equation (6.93), are

$$\alpha\begin{pmatrix} A & B \\ C & D \end{pmatrix}_t + \begin{pmatrix} -A & -B \\ C & D \end{pmatrix}_x + \begin{pmatrix} -A & B \\ -C & D \end{pmatrix}_z + 2\begin{pmatrix} -\hat{B}C & -\hat{B}D \\ \hat{C}A & \hat{C}B \end{pmatrix} = 0 \qquad (6.100)$$

and

$$\begin{pmatrix} \tilde{I}A & \tilde{I}B \\ 0 & 0 \end{pmatrix}_x + \begin{pmatrix} A\tilde{I} & 0 \\ C\tilde{I} & 0 \end{pmatrix}_z + \begin{pmatrix} [\tilde{I},\hat{A}]A+vC & [\tilde{I},A]B+vD \\ vA & vB \end{pmatrix} = 0 \qquad (6.101)$$

both for $z > x$. Equation (6.101) would seem to imply that $B = 0$, for $z > x$, which in turn makes it impossible to generate solutions of the SG equation. However $B \neq 0$ if the four-vector, ψ, on which equation (6.101) operates is not comprised of four linearly independent components. Indeed, it can be shown that the SG problem is *degenerate* (cf. §6.1.5(b)) in that, if

$$\psi = \begin{pmatrix} \psi_1 \\ \psi_2 \end{pmatrix},$$

where ψ_1, ψ_2 are two-vectors, then ψ_1 and ψ_2 are linearly dependent. (This property is discussed in Q6.15.) The construction of solutions of the SG equation, using the ZS scheme, is therefore far from straightforward even though the equations for F,

$$\begin{pmatrix} \tilde{I} & 0 \\ 0 & 0 \end{pmatrix} F_x + F_z \begin{pmatrix} \tilde{I} & 0 \\ 0 & 0 \end{pmatrix} = 0,$$

$$\alpha F_t + \begin{pmatrix} -I & 0 \\ 0 & I \end{pmatrix} F_x + F_z \begin{pmatrix} -I & 0 \\ 0 & I \end{pmatrix} = 0,$$

look unremarkable.

6.3 Two examples

We end with an outline of how to construct solutions of the NLS and sine–Gordon equations. Usually the most convenient method is to work with the ZS scheme, thus avoiding any specific calculations of the scattering data. However, the difficulties associated with the SG problem (mentioned above) suggest that we should use the AKNS scheme in this case. We shall therefore take the opportunity to give one example of each scheme: first the NLS equation using the ZS scheme, and second the sine–Gordon equation using the AKNS scheme. We shall describe the solitary-wave solutions of these equations, but we shall also mention the generalisations necessary to produce the N-soliton solutions.

Example (i): the nonlinear Schrödinger equation

The nonlinear Schrödinger equation, (6.90), is

$$i\alpha(l - m)u_t + (l + m)u_{xx} \pm \frac{2}{lm}(l - m)(l^2 - m^2)u|u|^2 = 0 \qquad (6.102)$$

which we shall simplify by making a suitable choice of l, m and α (see later) so as to enable us to solve

$$iu_t + u_{xx} + u|u|^2 = 0; \qquad (6.103)$$

this is the equation discussed in Q2.3. From §6.2.4(a) we have the equations for $F(x, z; t)$ as

$$\begin{pmatrix} l & 0 \\ 0 & m \end{pmatrix} F_x + F_z \begin{pmatrix} l & 0 \\ 0 & m \end{pmatrix} = 0 \qquad (6.104)$$

and

$$i\alpha F_t + F_{zz} - F_{xx} = 0. \qquad (6.105)$$

If we write

$$F = \begin{pmatrix} 0 & r \\ s & 0 \end{pmatrix}$$

then equation (6.104) shows that

$$r = r(mx - lz; t) \qquad \text{and} \qquad s = s(lx - mz; t),$$

and from equation (6.105) it is clear that there is a simple exponential solution with

$$r(x, z; t) = r_0 \exp \{\rho(mx - lz) + i\rho^2(l^2 - m^2)t/\alpha\} \qquad (6.106)$$

and

$$s(x, z; t) = s_0 \exp \{\sigma(lx - mz) + i\sigma^2(m^2 - l^2)t/\alpha\}, \qquad (6.107)$$

where r_0, s_0, ρ and σ are arbitrary constants.

The equation for $K_+(x, z; t)$ is the Marchenko equation

$$K_+(x, z; t) + F(x, z; t) + \int_x^\infty K_+(x, y; t)F(y, z; t)\,dy = 0$$

(see equation (6.67)) where we write

$$K_+ = \begin{pmatrix} a & b \\ c & d \end{pmatrix}$$

to give

$$\begin{pmatrix} a & b \\ c & d \end{pmatrix} + \begin{pmatrix} 0 & r \\ s & 0 \end{pmatrix} + \int_x^\infty \begin{pmatrix} a & b \\ c & d \end{pmatrix} \begin{pmatrix} 0 & r \\ s & 0 \end{pmatrix} dy = 0$$

(upon the suppression of the arguments of the functions). Thus, we obtain the scalar integral equations

$$a(x, z; t) + \int_x^\infty b(x, y; t)s_0 \exp \{\sigma(ly - mz) + i\sigma^2(m^2 - l^2)t/\alpha\}\,dy = 0$$

and

$$b(x, z; t) + r_0 \exp \{\rho(mx - lz) + i\rho^2(l^2 - m^2)t/\alpha\}$$
$$+ \int_x^\infty a(x, y; t)r_0 \exp \{\rho(my - lz) + i\rho^2(l^2 - m^2)t/\alpha\}\,dy = 0,$$

with two similar equations for $c(x, z; t)$ and $d(x, z; t)$. It now follows that

$$a(x, z; t) = e^{-\sigma mz}L(x, t), \qquad b(x, z; t) = e^{-\rho lz}M(x, t)$$

and thus the above two integral equations become

$$L + s_0 M \int_x^\infty \exp \{l(\sigma - \rho)y + i\sigma^2(m^2 - l^2)t/\alpha\}\,dy = 0$$

and

$$M + r_0 \exp\{\rho m x + i\rho^2(l^2 - m^2)t/\alpha\}$$
$$+ r_0 L \int_x^\infty \exp\{m(\rho - \sigma)y + i\rho^2(l^2 - m^2)t/\alpha\}\,dy = 0.$$

The integrals in these equations are defined provided

$$\mathscr{R}\{l(\sigma - \rho)\} < 0 \qquad \text{and} \qquad \mathscr{R}\{m(\rho - \sigma)\} < 0,$$

and so l and m must be of opposite sign. Let us choose, for example,

$$l = 2, \qquad m = -1 \qquad \text{and} \qquad \alpha = \tfrac{1}{3},$$

so that equation (6.102), with the *lower* sign, becomes

$$iu_t + u_{xx} + 9u|u|^2 = 0. \tag{6.108}$$

Hence, provided $\mathscr{R}(\sigma - \rho) < 0$, we obtain the solution for M,

$$M(x,t) = \frac{r_0 \exp(-\rho x + 9i\rho^2 t)}{\tfrac{1}{2}r_0 s_0(\rho - \sigma)^{-2} \exp\{3(\sigma - \rho)(x - 3i(\rho + \sigma)t)\} - 1}$$

and then

$$b(x,x;t) = \frac{r_0 \exp\{-3\rho(x - 3i\rho t)\}}{\tfrac{1}{2}r_0 s_0(\rho - \sigma)^{-2} \exp\{3(\sigma - \rho)(x - 3i(\rho + \sigma)t)\} - 1}. \tag{6.109}$$

From §6.2.4(a) we note that the solution of equation (6.108) is therefore

$$u(x,t) = b(x,x;t),$$

which is more conveniently expressed by choosing

$$\tfrac{1}{2}r_0 s_0(\rho - \sigma)^{-2} = -1,$$

with $\rho = k + i\lambda$, $\sigma = -k + i\lambda$ for $k > 0$. The solution given in equation (6.109) now simplifies to become

$$b(x,x;t) = -\tfrac{1}{2}r_0 \exp\{-3i\lambda x + 9i(k^2 - \lambda^2)t\}\operatorname{sech}(3kx + 18k\lambda t)$$

with

$$r_0 s_0 = -8k^2.$$

Finally, the corresponding analysis for the two functions $c(x,z;t)$ and $d(x,z;t)$, and the requirement that $c(x,x;t) = -u^*(x,t)$ (see §6.2.4(a)), implies that $s_0 = -r_0$ and so $r_0 = \pm 2\sqrt{2}k$. If we compare equations (6.103) and (6.108) we see that a solution of equation (6.103) is $3b(x,x;t)$; let us therefore introduce

$$a = 3\sqrt{2}k \qquad \text{and} \qquad c = -6\lambda.$$

A solution to our original NLS equation, (6.103), which represents the solitary wave, is thus

$$u(x,t) = \pm a \exp\left[i\left\{\frac{c}{2}(x - ct) + nt\right\} \right] \operatorname{sech}\{a(x - ct)/\sqrt{2}\}$$

where $n = \frac{1}{2}(a^2 + \frac{1}{2}c^2)$. This is the solution already given in Q2.3 (where there is no restriction that $a > 0$).

The generalisation of this method of solution to the construction of the N-soliton solution is immediate: the solutions given in equations (6.106) and (6.107) are replaced by sums of exponential terms of this type, e.g.

$$r(x, z; t) = \sum_{n=1}^{N} r_n \exp\{\rho_n(mx - lz) + i\rho_n^2(l^2 - m^2)t/\alpha\};$$

see Q6.16.

Example (ii): the sine–Gordon equation

The sine–Gordon (SG) equation written in the form

$$u_{xt} = \sin u \tag{6.110}$$

(see equations (6.97) and (6.99)) can be solved by using the AKNS scheme described in §6.1. Let us first summarise the details of this method as it applies to the SG equation.

The eigenvector, ψ, satisfies

$$\psi_t = \frac{1}{\zeta}\begin{pmatrix} A & B \\ B & -A \end{pmatrix}\psi, \qquad \psi_x = \begin{pmatrix} -i\zeta & q \\ -q & i\zeta \end{pmatrix}\psi,$$

where

$$A = \tfrac{1}{4}i\cos u, \qquad B = \tfrac{1}{4}i\sin u, \qquad q = -\tfrac{1}{2}u_x; \tag{6.111}$$

see equations (6.39) and §6.1.5(b). Then

$$q(x, t) = -2K_{12}(x, x; t) \tag{6.112}$$

where

$$K_{11}(x, z; t) + \int_x^\infty K_{12}(x, y; t)f(y + z; t)\,dy = 0 \tag{6.113}$$

and

$$K_{12}(x, z; t) - f(x + z; t) - \int_x^\infty K_{11}(x, y; t)f(y + z; t)\,dy = 0 \tag{6.114}$$

(see equations (6.35) and (6.36)) with

$$f(X; t) = \frac{1}{2\pi}\int_{-\infty}^\infty \frac{b_0(k)}{a(k)}\exp\{i(kX - \tfrac{1}{2}t/k)\}\,dk - i\sum_{n=1}^{N} c_n \exp\{i(\zeta_n X - \tfrac{1}{2}t/\zeta_n)\};$$

$$\tag{6.115}$$

see equations (6.26), (6.46) and (6.47). (Remember that

$$A_{11} = \frac{A}{\zeta} \rightarrow \frac{i}{4\zeta} \qquad \text{as } u \rightarrow 0$$

for the SG equation: see §6.1.5(b).) Here $b_0(k)$ and $a(k)$ are the appropriate scattering data, and c_n, $n = 1,\ldots,N$, are the normalisation constants at each discrete eigenvalue, $\zeta = \zeta_n$, all at $t = 0$; see equations (6.7) and (6.26).

The solitary-wave solution of the SG equation can be obtained by choosing

$$q(x, 0) = \lambda \operatorname{sech}(\lambda x + \mu);$$

this example was discussed in §6.1.3, although here we have allowed an arbitrary phase shift (μ). From §6.1.3 we find that

$$a(\zeta) = \frac{2i\zeta + \lambda}{2i\zeta - \lambda}$$

and

$$b_0(\zeta) = 0, \qquad \zeta \neq \tfrac{1}{2}i\lambda; \qquad b_0(\tfrac{1}{2}i\lambda) = -e^{-\mu},$$

so that there is only one discrete eigenvalue (at $\zeta = \zeta_1 = \tfrac{1}{2}i\lambda$, $\lambda > 0$). Thus

$$c_1 = -i\lambda e^{-\mu}$$

and then from equation (6.115) we obtain

$$f(X; t) = -\lambda e^{-(\lambda X/2 + t/\lambda + \mu)}.$$

The pair of coupled integral equations, (6.113) and (6.114), now become

$$K_{11} - \lambda e^{-(\mu + t/\lambda)} \int_x^\infty K_{12}(x, y; t) e^{-\lambda(y + z)/2} \, dy = 0 \qquad (6.116)$$

and

$$K_{12} + \lambda e^{-(\mu + t/\lambda) - \lambda(x + z)/2} + \lambda e^{-(\mu + t/\lambda)} \int_x^\infty K_{11}(x, y; t) e^{-\lambda(y + z)/2} \, dy = 0,$$

which imply

$$K_{11}(x, z; t) = L(x, t) e^{-\lambda z/2}.$$

Hence

$$K_{12} + \{\lambda e^{-(\mu + t/\lambda + \lambda x/2)} + L e^{-(\mu + t/\lambda + \lambda x)}\} e^{-\lambda z/2} = 0$$

and so from equation (6.166) we obtain

$$L + (\lambda e^{-\lambda x/2} + L e^{-\lambda x}) e^{-2(\mu + t/\lambda) - \lambda x} = 0.$$

Thus

$$L(x, t) = -\frac{\lambda e^{-2(\mu + t/\lambda) - 3\lambda x/2}}{1 + e^{-2(\mu + t/\lambda + \lambda x)}}$$

which gives

$$K_{12}(x, z; t) = -\frac{\lambda e^{-(\mu + t/\lambda) - \lambda(x+z)/2}}{1 + e^{-2(\mu + t/\lambda + \lambda x)}},$$

and then from equations (6.111) and (6.112) we have

$$u_x(x, t) = -\frac{4\lambda e^{-(\mu + \lambda x + t/\lambda)}}{1 + e^{-2(\mu + \lambda x + t/\lambda)}}.$$

This equation is integrated once to give

$$u(x, t) = -4 \arctan(e^{\mu + \lambda x + t/\lambda}),$$

if $u \to 0$ as $x \to -\infty$, which is conveniently written as

$$u(x, t) = 4 \arctan(ce^{\lambda x + t/\lambda}), \qquad (6.117)$$

where $c = -e^{\mu}$. This is the solitary-wave solution of the sine–Gordon equation (6.110), for arbitrary constants c and $\lambda \neq 0$ (see equation (5.71)). (Although our derivation has been for $\lambda > 0$, solution (6.117) is also valid for $\lambda < 0$ essentially because $f = \hat{f}$: see §6.1.3). The solution with $\lambda > 0$ is usually called a 'kink', and with $\lambda < 0$ it is an 'antikink'; see §8.1.

The generalisation to N-solitons is most easily accomplished by expressing $f(X; t)$ as a sum of exponential terms of the type used above, i.e.

$$f(X; t) = -\sum_{n=1}^{N} \lambda_n \exp\{-(\tfrac{1}{2}\lambda_n X + t/\lambda_n + \mu_n)\};$$

see exercise Q6.17.

Further reading

6.1 The AKNS scheme is described extensively in Ablowitz & Segur (1981) and Lamb (1980).

6.2 The ZS scheme is described in the paper by Zakharov & Shabat (1974) and also in Appendix 3 of Novikov, Manakov, Pitavskii & Zakharov (1984). Elements from both schemes are presented together in Dodd, Eilbeck, Gibbon & Morris (1982, Chap. 6), and Newell (1985, §3(c)).

Exercises

Q6.1 *Symmetries of the eigenfunctions.*

(i) Show that, if $r = \pm q$ is real and $\psi = \begin{pmatrix} \psi_1(x; \zeta) \\ \psi_2(x; \zeta) \end{pmatrix}$ is a solution of the pair of equations (6.1) and (6.2), then so is

$$\psi = \begin{pmatrix} \psi_2(x; -\zeta) \\ \pm \psi_1(x; -\zeta) \end{pmatrix}.$$

(ii) Show that, if $r = \pm q^*$, where the asterisk denotes the complex conjugate and $\psi = \begin{pmatrix} \psi_1(x;\zeta) \\ \psi_2(x;\zeta) \end{pmatrix}$ is a solution as above, then so is

$$\psi = \begin{pmatrix} \psi_2^*(x;\zeta^*) \\ \pm \psi_1^*(x;\zeta^*) \end{pmatrix}.$$

Q6.2 *Wronskian relations.* Show that, if $\begin{pmatrix} \psi_1 \\ \psi_2 \end{pmatrix}$ and $\begin{pmatrix} \phi_1 \\ \phi_2 \end{pmatrix}$ are two solutions of the pair of equations (6.1) and (6.2), then

$$\frac{\partial}{\partial x} W(\psi,\phi) = 0,$$

where

$$W(\psi,\phi) = \psi_1\phi_2 - \psi_2\phi_1.$$

Hence deduce that

$$W(\psi_+,\hat{\psi}_+) = 1 \qquad \text{and} \qquad W(\psi_-,\hat{\psi}_-) = -1,$$

where $\psi_+, \hat{\psi}_+, \psi_-$ and $\hat{\psi}_-$ are defined by equations (6.4) and (6.5).
Also show that

$$W(\psi_-,\hat{\psi}_+) = a(\zeta),$$

and obtain similar expressions for $W(\psi_-,\psi_+)$, $W(\hat{\psi}_-,\psi_+)$ and $W(\hat{\psi}_-,\hat{\psi}_+)$.
Finally, by considering the form of $W(\psi_-,\hat{\psi}_-)$ as $x \to +\infty$, obtain the law of energy conservation in the form

$$a\hat{a} + b\hat{b} = 1.$$

Q6.3 *Symmetries of the scattering data.* Use the results of Q6.1 and Q6.2 to show that, if $r = \pm q$ (real), then

$$\begin{pmatrix} \hat{\psi}_{1+}(x;\zeta) \\ \hat{\psi}_{2+}(x;\zeta) \end{pmatrix} = \begin{pmatrix} \pm\psi_{2+}(x;-\zeta) \\ \psi_{1+}(x;-\zeta) \end{pmatrix}$$

and

$$\begin{pmatrix} \hat{\psi}_{1-}(x;\zeta) \\ \hat{\psi}_{2-}(x;\zeta) \end{pmatrix} = \begin{pmatrix} \mp\psi_{2-}(x;-\zeta) \\ -\psi_{1-}(x;\zeta) \end{pmatrix},$$

and hence deduce that

$$\hat{a}(\zeta) = a(-\zeta), \qquad \hat{b}(\zeta) = \mp b(-\zeta).$$

Now obtain the corresponding results if $r = \pm q^*$.

Q6.4 *Symmetries of the Marchenko equation.* Use the results of Q6.3 to show that, if $r = \pm q$ (real), then

$$K_{21} = \pm K_{12} \qquad \text{and} \qquad K_{22} = K_{11},$$

where

$$K_{11} + \int_x^\infty K_{12} f \, dy = 0; \qquad \pm K_{12} + f + \int_x^\infty K_{11} f \, dy = 0$$

(see equation (6.24)).
Now obtain the corresponding results if $r = \pm q^*$.

Q6.5 *Conserved quantities.*
(i) Eliminate ψ_2 between the pair of equations

$$\psi_{1x} = -i\zeta\psi_1 + q\psi_2; \qquad \psi_{2x} = i\zeta\psi_2 + r\psi_1,$$

and then with the choice $\psi_1 = \psi_{1-}$ let $\psi_{1-} = e^{-i\zeta x + \theta}$ and obtain the equation for $\theta(x; \zeta, t)$.

(ii) Now, since $\theta \to 0$ as $|\zeta| \to \infty$, assume that

$$\theta_x(x; \zeta, t) = \sum_{n=0}^{\infty} \frac{c_n(x, t)}{(2i\zeta)^{n+1}}$$

and hence show that

$$c_0 = -qr, \qquad c_1 = -qr_x;$$

$$c_{n+1} = q\left(\frac{c_n}{q}\right)_x + \sum_{m=0}^{n-1} c_m c_{n-m-1} \quad (n \geqslant 1).$$

(iii) Deduce that

$$\lim_{x \to \infty} (e^{i\zeta x} \psi_{1-}) = a(\zeta)$$

and hence show that

$$\log a(\zeta) = \sum_{n=0}^{\infty} \frac{1}{(2i\zeta)^{n+1}} \int_{-\infty}^{\infty} c_n(x, t)\, dx$$

and thus obtain the conserved quantities

$$\int_{-\infty}^{\infty} c_n(x, t)\, dx = \text{constant}, \qquad n \geqslant 0.$$

(Zakharov & Shabat, 1972)

Q6.6 *Conserved quantities again.* Use the results of Q6.5 to obtain the first three conserved quantities in the two cases: (i) $r = \pm q$ (real), and (ii) $r = \pm q^*$ (see Q5.5).

Q6.7 *Modified KdV equation.* Show that the choice

$$A_{11} = -A_{22} = c\{\zeta^3 + \tfrac{1}{2}qr\zeta - \tfrac{1}{4}i(qr_x - q_x r)\};$$

$$A_{12}(q, r, \zeta) = c(iq\zeta^2 - \tfrac{1}{2}q_x\zeta + \tfrac{1}{2}iq^2 r - \tfrac{1}{4}iq_{xx});$$

$$A_{21} = A_{12}(r, q, -\zeta),$$

(see §6.1.5), with $r = \pm q$ and c a suitably chosen constant, gives the mKdV equations

$$q_t \mp 6q^2 q_x + q_{xxx} = 0.$$

Q6.8 *NLS equation.* Show that the equation

$$iq_t + q_{xx} - q|q|^2 = 0$$

(see Q2.4), has a solitary-wave solution for which $|q| \nrightarrow 0$ as $|x| \to \infty$.

Q6.9 *Sinh–Gordon equation.* Show that the choice

$$A_{11} = -A_{22} = \frac{i}{4\zeta}\cosh u, \qquad A_{12} = -A_{21} = -\frac{i}{4\zeta}\sinh u,$$

with $q = r = \frac{1}{2}u_x$ (cf. §6.1.5(b)), gives

$$u_{xt} = \sinh u.$$

Q6.10 *Operator* K_-. Deduce from equation (6.66) the identity

$$K_-(x,z) = F(x,z) + \int_x^\infty K_+(x,y)F(y,z)\,dy.$$

Q6.11 *Operator identity.* Show that, if

$$\Delta(I + J_+) = (I + J_+)\Delta_0 \qquad \text{and} \qquad (I + J_+)(I + J_F) = I + J_-,$$

where Δ_0 and J_F commute, then

$$\Delta(I + J_-) = (I + J_-)\Delta_0.$$

Q6.12 *Commuting operators.* Verify that the following pairs of operators commute:

(i) $\Delta_0^{(1)} = \alpha\dfrac{\partial}{\partial t} + \displaystyle\sum_{n=1}^{N}\alpha_n\dfrac{\partial^n}{\partial x^n}$, $\qquad \Delta_0^{(2)} = \beta\dfrac{\partial}{\partial y} + \displaystyle\sum_{m=1}^{M}\beta_m\dfrac{\partial^m}{\partial x^m}$,

where α, α_n, β and β_m are constant scalars;

(ii) $\Delta_0^{(1)} = \begin{pmatrix} 1 & 0 \\ 0 & 1 \end{pmatrix}\left(i\alpha\dfrac{\partial}{\partial t} - \dfrac{\partial^2}{\partial x^2}\right)$, $\qquad \Delta_0^{(2)} = \begin{pmatrix} l & 0 \\ 0 & m \end{pmatrix}\dfrac{\partial}{\partial x}$,

where α, l and m are constants;

(iii) $\Delta_0^{(1)} = \begin{pmatrix} I & 0 \\ 0 & I \end{pmatrix}\alpha\dfrac{\partial}{\partial t} + \begin{pmatrix} -I & 0 \\ 0 & I \end{pmatrix}\dfrac{\partial}{\partial x}$, $\qquad \Delta_0^{(2)} = \begin{pmatrix} \tilde{I} & 0 \\ 0 & 0 \end{pmatrix}\dfrac{\partial}{\partial x}$,

where α is a constant.

Q6.13 *KdV equation: operators.* Show that, if

$$\Delta_0^{(1)} = \frac{\partial}{\partial t} + 4\frac{\partial^3}{\partial x^3},$$

then $\Delta_0^{(1)}$ and J_F commute provided

$$F_t + 4(F_{xxx} + F_{zzz}) = 0;$$

see equation (6.84).

Q6.14 *Two-dimensional KdV equation.* If

$$\Delta_0^{(1)} = \frac{\partial}{\partial t} + 4\frac{\partial^3}{\partial x^3}, \qquad \Delta_0^{(2)} = \frac{\partial}{\partial y} - \frac{\partial^2}{\partial x^2},$$

and we write

$$\Delta^{(2)} = \Delta_0^{(2)} + u(x,t,y),$$

show that

$$\Delta^{(1)} = \Delta_0^{(1)} - 6u\frac{\partial}{\partial x} - 3u_x + w(x, t, y).$$

Hence deduce that

$$u_t - 6uu_x + u_{xxx} - w_y = 0$$

with

$$w_x = -3u_y,$$

which together give the 2D KdV (or Kadomtsev–Petviashvili) equation.

Q.6.15 *Sine–Gordon equation.* Write

$$\psi_{1x} = R'\psi_1 + i\zeta\begin{pmatrix} -1 & 0 \\ 0 & 1 \end{pmatrix}\psi_1, \qquad \psi_{1t} = \frac{1}{\zeta}A'\psi_1,$$

where R' and A' and 2×2 matrices independent of ζ (see §6.1.5(b)) and introduce the two-vector ψ_2 defined by

$$B\psi_1 = \zeta\psi_2.$$

Hence obtain the two equations

$$L\psi_x + M\psi = \zeta N\psi, \qquad L'\psi_t + M'\psi = \zeta N'\psi$$

for suitable 4×4 matrices L, L', M, M' and N, N' independent of ζ (where L, L', N, N' are constant matrices), operating on the four-vector $\psi = \begin{pmatrix} \psi_1 \\ \psi_2 \end{pmatrix}$.

Q6.16 *NLS equation: soliton solution.* Obtain the two-soliton solution of the NLS equation

$$iu_t + u_{xx} + u|u|^2 = 0;$$

see §6.3, example (i).

Q6.17 *Sine–Gordon equation: soliton solution.* Obtain the two-soliton solution of the SG equation

$$u_{xt} = \sin u$$

in the form

$$u(x, t) = 4\arctan\left\{\left(\frac{\lambda_1 + \lambda_2}{\lambda_1 - \lambda_2}\right)\frac{\exp(\theta_1) - \exp(\theta_2)}{1 + \exp(\theta_1 + \theta_2)}\right\},$$

where $\theta_i = \lambda_i x + t/\lambda_i + \mu_i$; see §6.3, example (ii).

Q6.18 *Sinh–Gordon equation.* Use the method of §6.3, example (ii), and Q6.9 to obtain the solitary-wave solution of the equation

$$u_{xt} = \sinh u.$$

Q6.19 *Modified KdV equation.* Obtain the two-soliton solution of the mKdV equation

$$u_t + 6u^2u_x + u_{xxx} = 0;$$

see Q6.7. Also write down the solitary-wave solution of this equation.

Q6.20 *Modified KdV equation: breather solution.* From the two-soliton solution of the mKdV equation (see Q6.19), obtain the breather solution by choosing the two eigenvalues to satisfy $\zeta_2 = -\zeta_1^*$.

Q6.21 *Sine–Gordon equation; breather solution.* Let $a_1 = \alpha + i\beta$ and choose $a_2 = a_1^*$, and hence obtain the breather solution of the sine–Gordon equation

$$u_{xt} = \sin u$$

from the two-soliton solution given in Q6.17.

Q6.22 *Sine–Gordon solution: asymptotic behaviour.* Describe the behaviour of the two-soliton solution of the sine–Gordon equation, given in Q6.17, as $t \to -\infty$; for $t = 0$; as $t \to +\infty$.

[You may find it more convenient to work with u_x rather than u itself.]

Q6.23 *Kinks and antikinks.* Take the two-soliton solution of the sine–Gordon equation, given in Q6.17, and find the conditions necessary for
 (i) a kink–kink interaction;
 (ii) a kink–antikink interaction.

Q6.24 *Davey–Stewartson equations.* Show that the choice

$$\Delta_0^{(1)} = I\left(\alpha\frac{\partial}{\partial t} + \frac{\partial^2}{\partial x^2}\right); \qquad \Delta_0^{(2)} = I\beta\frac{\partial}{\partial y} + i\begin{pmatrix} 1 & 0 \\ 0 & -1 \end{pmatrix}\frac{\partial}{\partial x},$$

leads to

$$\Delta^{(1)} = \Delta_0^{(1)} - \begin{pmatrix} \phi & \psi_x \\ \psi_x^* & \phi^* \end{pmatrix}; \qquad \Delta^{(2)} = \Delta_0^{(2)} - i\begin{pmatrix} 0 & \psi \\ -\psi^* & 0 \end{pmatrix},$$

where $\phi = \phi(x, y, t)$ and $\psi = \psi(x, y, t)$.

Hence obtain the Lax equation for this system in the form of a pair of equations

$$i\alpha\psi_t = \beta\psi_{xy} + i\psi(\phi' - \phi^*); \qquad \beta\phi_y + i\phi_x = i|\psi|^2.$$

[This system arises in the study of nearly monochromatic, nearly one-dimensional wave packets on the surface of water: see Davey & Stewartson (1974), Anker & Freeman (1978a).]

Q6.25 *Boussinesq equation.* Show that the choice

$$\Delta_0^{(1)} = \frac{i}{\sqrt{3}}\frac{\partial}{\partial t} - \frac{\partial^2}{\partial x^2}, \qquad \Delta_0^{(2)} = \frac{\partial^3}{\partial x^3} + \frac{\partial}{\partial x}$$

leads to

$$\Delta^{(1)} = \Delta_0^{(1)} + u(x, t); \qquad \Delta^{(2)} = \Delta_0^{(2)} - 6u\frac{\partial}{\partial x} - 3u_x + w(x, t).$$

Hence show that

$$w_x = -i\sqrt{3}\,u_t$$

and then

$$u_{tt} - u_{xx} + 3(u^2)_{xx} - u_{xxxx} = 0,$$

the Boussinesq equation; see Q5.21 and Hirota (1973b).

What pair of equations does $F(x, t)$ satisfy for the above operators?

[This Boussinesq equation admits solutions which describe *head-on* collisions between solitons.]

*The Painlevé property, perturbations and numerical methods

This chapter is devoted to three additional topics.

The equations we have discussed have been those for which the inverse scattering transform is applicable. However, given *any* evolution equation, it is natural to ask whether it can be solved by the inverse scattering transform (IST); in other words, how do we decide if a given equation is *completely integrable*? This question is still open but a promising conjecture concerns the so-called *Painlevé property*. We shall describe how the Painlevé equations arise, what they are and the conjecture itself.

If the evolution equation cannot be solved by the IST, but is close to one which can be (by virtue of a small parameter), we may adopt the following procedure. The IST method is formulated in the conventional way but the time evolution of the scattering data now involves the small parameter. This parameter can be used as the basis for generating an asymptotic solution of the inverse scattering problem, and hence of the original equation. We shall outline the development of this argument.

Finally, if neither of the above methods is applicable, or if a graphical representation of the solution is required, then we may use a numerical solution. Indeed, the original motivation for the IST came from a study of numerical solutions of the KdV equation. We shall, in the final section of this chapter, present some numerical methods suited to the solution of the initial-value problem for evolution equations.

7.1 The Painlevé property

7.1.1 Painlevé equations

Solutions of ordinary differential equations may incorporate singularities of one sort or another. So, for example, the solution might have a pole or a branch point; note that we are thinking of solutions in the complex plane. Now the singularities of the solutions of linear differential equations are always at points which are *fixed*, i.e. their positions are independent of the arbitrary constants of integration. A simple example of this type is the equation

$$z\frac{dw}{dz} + w = 0,$$

which has the general solution

$$w(z) = c/z,$$

where c is an arbitrary constant. This solution has a simple pole at $z = 0$, for all $c \neq 0$.

Nonlinear ordinary differential equations, on the other hand, may have solutions which have *movable* singularities, i.e. whose position does depend on the arbitrary constants of integration. Thus, for example, the equation

$$\frac{dw}{dz} + w^2 = 0$$

has the general solution

$$w(z) = \frac{1}{z - z_0},$$

where z_0 is an arbitrary complex number; this solution has a simple pole at the movable point $z = z_0$.

It is convenient to differentiate between poles and all other singularities of an ordinary differential equation. A *critical point* is a singularity, i.e. a point at which the solution is not analytic, which is *not* a pole. Thus a critical point might be a branch point or an essential singularity; for example, the equation

$$\frac{dw}{dz} = e^{-w}$$

has the general solution

$$w(z) = \log(z - z_0),$$

where z_0 is an arbitrary constant. This solution therefore has a logarithmic branch point at the movable point $z = z_0$.

Now, we are concerned here only with those equations which do *not* contain movable critical points. Such equations have long been of interest to mathematicians. Indeed, in 1884, Fuchs (see Ince, 1927) showed that if the first order equation

$$\frac{dw}{dz} = F(z, w), \tag{7.1}$$

where F is rational in w and analytic in z, does not contain any movable critical points then

$$\frac{dw}{dz} = F(z, w) = a(z) + b(z)w + c(z)w^2, \tag{7.2}$$

for some analytic functions a, b and c; this is a generalised Riccati equation. At the turn of this century, Painlevé and Gambier extended these ideas to equations of the second order,

$$\frac{d^2w}{dz^2} = F\left(z, w, \frac{dw}{dz}\right). \tag{7.3}$$

They found that, similarly, if F is rational in w and dw/dz, and analytic in z, then there are 50 different cases. Of these 50 equations, all but six could be solved in finite terms of known functions, i.e. elementary functions or elliptic functions; the other six equations have solutions which are called *Painlevé transcendents*. They are usually labelled P-I to P-VI, the first three being

P-I: $$\frac{d^2w}{dz^2} = 6w^2 + z; \tag{7.4}$$

P-II: $$\frac{d^2w}{dz^2} = zw + 2w^3 + \alpha; \tag{7.5}$$

P-III: $$\frac{d^2w}{dz^2} = \frac{1}{w}\left(\frac{dw}{dz}\right)^2 - \frac{1}{z}\frac{dw}{dz} + \frac{1}{z}(\alpha w^2 + \beta) + \gamma w^3 + \frac{\delta}{w}, \tag{7.6}$$

where α, β, γ and δ are arbitrary constants. These equations cannot be reduced to simpler equations: they are *irreducible*. (A complete description of these first and second order equations is given in Ince (1927, Chaps. 13 and 14).)

7.1.2 The Painlevé conjecture

We now turn to the relation between evolution equations which can be solved by the inverse scattering transform, and the Painlevé equations. First, for convenience, we refer to the absence of movable critical points for an ordinary differential equation as the *Painlevé property*. Thus the 50 equations mentioned in §7.1.1 are the only rational second order equations which satisfy the Painlevé property.

Before we discuss evolution equations in general, let us recall Q1.13. This shows that the KdV equation

$$u_t - 6uu_x + u_{xxx} = 0$$

has the similarity solution

$$u(x, t) = -(3t)^{-2/3}F(\eta), \qquad \eta = x(3t)^{-1/3},$$

where

$$F''' + (6F - \eta)F' - 2F = 0.$$

Then if $F = \lambda\, dV/d\eta - V^2$ (for some λ) it follows that

$$V'' = \eta V + 2V^3, \qquad (7.7)$$

provided V decays exponentially as either $\eta \to +\infty$ or $\eta \to -\infty$. We see immediately that the equation for $V(\eta)$ is the P-II equation with $\alpha = 0$. Thus, the KdV equation can be reduced to an associated ordinary differential equation which has the Painlevé property. This, and numerous similar results, have suggested the *Painlevé conjecture*, first formulated by Ablowitz, Ramani & Segur (1978; also 1980a, b) and also studied by McLeod & Olver (1983). This can be expressed as:

A nonlinear partial differential equation is solvable by the inverse scattering transform if, and only if, every ordinary differential equation derived from it (by exact reduction) satisfies the Painlevé property.

Such an ordinary differential equation may result, for example, from a search for a wave of permanent form or for a similarity solution. Also, it may require a further transformation of variables to put it into a standard Painlevé form. This is no more than a conjecture: its proof, or disproof, is still awaited. However, much evidence and many convincing arguments have been given recently so that we may regard the conjecture as a practical test for the existence of an inverse scattering transform. (The Painlevé property has also been related directly to the evolution equations; see Weiss, Tabor & Carnevale, 1983.)

Reductions of the KdV, modified KdV, concentric KdV and NLS equations have already been met: see Q1.13, 2.9, 2.10, 2.11, respectively. Other examples will be found in the exercises at the end of this chapter.

7.1.3 *Linearisation of the Painlevé equations*

The relation between the inverse scattering transform and the Painlevé equations now leads us to a novel result. Since the evolution equations can be linearised via the Marchenko integral equation, and a special reduction of an evolution equation is expected to be a Painlevé equation, we should be able to linearise the Painlevé equation itself. In the case of the equations P-I to P-VI this may be a surprise: they can not be solved in terms of any known functions, indeed they define new functions (the Painlevé transcendents mentioned above).

Furthermore, solutions of the various evolution equations we have discussed are related by Bäcklund transformations. Thus we anticipate that solutions of, for example, the P-I to P-VI equations are related in the same way. This particular property has been exploited by Airault (1979) to obtain rational solutions of certain equations of the types P-II, P-III, P-IV and P-V.

We end these comments by showing how a particular Painlevé equation (P-II, i.e. equation (7.5)) can be linearised. From Q4.2 we know that the concentric KdV equation,

$$u_t + \frac{u}{2t} + 6uu_x + u_{xxx} = 0 \qquad (7.8)$$

has the solution

$$u(x,t) = \frac{2}{(12t)^{2/3}} \frac{\partial}{\partial X} K(X, X; t), \qquad (7.9)$$

where $X = x/(12t)^{1/3}$ and $K(x, z; t)$ is the solution of

$$K(x, z; t) + F(x, z; t) + \int_x^\infty K(x, y; t) F(y, z; t)\, dy = 0. \qquad (7.10)$$

The function F satisfies the pair of equations

$$F_{xx} - F_{zz} = (x - z)F,$$
$$3tF_t - F + F_{xxx} + F_{zzz} = xF_x + zF_z,$$

which has the solution

$$F(x, z; t) = \int_{-\infty}^\infty f(st^{1/3}) \mathrm{Ai}(x + s) \mathrm{Ai}(z + s)\, ds, \qquad (7.11)$$

where f is an arbitrary function – possibly a generalised function – and Ai is the Airy function.

Furthermore, if we follow the derivation given in Q2.10, we can show that equation (7.8) has a similarity solution

$$u(x,t) = -\frac{1}{(12t)^{2/3}} g\left(\frac{x}{(12t)^{1/3}}\right), \qquad (7.12)$$

where the equation for the function g can be expressed in the form

$$\frac{d^2 v}{dX^2} - Xv - 2v^3 = 0, \qquad X = x/(12t)^{1/3}, \qquad (7.13)$$

if $g = 2v^2$ and $g \to 0$ as $X \to +\infty$. (Remember that the inverse scattering transform method requires potential functions, u, which decay sufficiently rapidly at infinity.) Equation (7.13) is the P-II equation with $\alpha = 0$ (see equation (7.5)). If we compare solutions (7.9) and (7.12), we see that they will be the same solution if we are able to choose $K(X, X; t)$ to be a function only of X. This requires that F is not a function of t, and from equation (7.11) it follows that this is indeed possible. The relevant solution is obtained by setting $f(s) = kH(s)$, where H is Heaviside's step function and k is an arbitrary constant. From equation (7.11) we therefore obtain

$$F(x, z; t) = k \int_0^\infty \mathrm{Ai}(x + s)\mathrm{Ai}(z + s)\,\mathrm{d}s, \tag{7.14}$$

so that there is no t-dependence. Equation (7.10) now gives

$$L(x, s) + k\mathrm{Ai}(x + s) + k \int_0^\infty L(x, p) \int_x^\infty \mathrm{Ai}(p + q)\mathrm{Ai}(q + s)\,\mathrm{d}q\,\mathrm{d}p = 0 \tag{7.15}$$

where

$$K(x, z; t) = \int_0^\infty L(x, s)\mathrm{Ai}(z + s)\,\mathrm{d}s. \tag{7.16}$$

It is clear from equations (7.16), (7.9) and (7.12) that

$$v^2(X) = -2\frac{\mathrm{d}}{\mathrm{d}X} \int_0^\infty L(X, s)\mathrm{Ai}(X + s)\,\mathrm{d}s; \tag{7.17}$$

in other words, equations (7.15) and (7.17) constitute a linearisation of the Painlevé equation, (7.13). The constant k enables a one-parameter family of solutions to be generated, although the Neumann expansion obtained from equation (7.15) is convergent only if $0 \leqslant k < \frac{1}{2}$. Finally, it turns out that the solution, $v(X)$, can be expressed in terms of L in another way: it can be shown that

$$v(X) = \mathrm{i}\left(\frac{2}{k}\right)^{1/2} L(X, 0).$$

(The appearance of i here is not particularly significant: the transformation $v \to \mathrm{i}v$ merely changes the sign of the v^3 term in equation (7.13).) Further details concerning the linearisation of the Painlevé equations can be found in Ablowitz & Segur (1977), Johnson (1979) and Fokas & Ablowitz (1981).

7.2 Perturbation theory

The reconstruction of a function, $u(x, t)$, by the inverse scattering transform method is feasible when the discrete eigenvalues are constant and the scattering coefficients evolve in an elementary way. Nevertheless, the same procedure can be adopted even when the equation is not exactly integrable. If the terms which make a given equation non-integrable are all associated with a small parameter, ε say, then the derivation given in Chap. 4 can still be exploited. Now, however, the time evolution of the scattering data will depend on ε and it can therefore be described asymptotically as $\varepsilon \to 0$. We shall outline this argument for the perturbed KdV equation

$$u_t - 6uu_x + u_{xxx} = \varepsilon q(x, t; \varepsilon), \qquad (7.18)$$

where q may depend on x and t via $u(x, t)$ itself, and $q \to 0$ as $|x| \to \infty$. Indeed, there is a classical problem of this type where

$$q(x, t; \varepsilon) = \gamma(t; \varepsilon) u(x, t),$$

which describes the propagation of long waves in water of variable depth (Johnson, 1973).

The development follows closely that given in Chap. 4. Thus, from §4.3, for the time evolution of the scattering data, we have

$$\frac{\partial}{\partial x}(\psi_x R - \psi R_x) = \psi^2(\lambda_t - u_t + 6uu_x - u_{xxx})$$

where

$$R = \psi_t + u_x \psi - 2(u + 2\lambda)\psi_x.$$

Upon the use of equation (7.18) we obtain

$$\frac{\partial}{\partial x}(\psi_x R - \psi R_x) = \psi^2(\lambda_t - \varepsilon q) \qquad (7.19)$$

and so, if the discrete spectrum is $\lambda = -\kappa_n^2 \, (<0)$ with $\psi = \psi_n \, (n = 1, 2, \ldots, N)$, then

$$(\kappa_n^2)_t + \varepsilon \int_{-\infty}^{\infty} \psi_n^2 q \, dx = 0. \qquad (7.20)$$

Hence

$$\kappa_{nt} = O(\varepsilon) \qquad \text{as } \varepsilon \to 0,$$

at least for $t = O(1)$, and consequently the discrete eigenvalues will evolve *slowly* in t. It is convenient to write

$$\lambda_t = -(\kappa_n^2)_t = \varepsilon \omega_n, \qquad (7.21)$$

so that equation (7.19) becomes, for the discrete spectrum,

$$\frac{\partial}{\partial x}(\psi_{nx} R_n - \psi_n R_{nx}) = \varepsilon \psi_n^2 (\omega_n - q)$$

or

$$\psi_{nxx}R_n - \psi_n R_{nxx} = \varepsilon\psi_n^2(\omega_n - q).$$

If we now use the Sturm–Liouville equation for ψ_n,

$$\psi_{nxx} - (\kappa_n^2 + u)\psi_n = 0,$$

the equation for $R_n(x, t; \varepsilon)$ can be written as

$$R_{nxx} - (\kappa_n^2 + u)R_n = \varepsilon(q - \omega_n)\psi_n. \tag{7.22}$$

Correspondingly, for the continuous spectrum $\lambda = k^2 \ (>0)$, equation (7.19) gives

$$\hat{\psi}_{xx}\hat{R} - \hat{\psi}\hat{R}_{xx} = -\varepsilon q\hat{\psi}^2$$

if we follow the evolution at fixed k. Thus the equation for $\hat{R}$ becomes

$$\hat{R}_{xx} + (k^2 - u)\hat{R} = \varepsilon q\hat{\psi}. \tag{7.23}$$

It is convenient to express $\hat{R}$ as

$$\hat{R} = A(k)\hat{\psi} + \varepsilon\hat{r}, \tag{7.24}$$

where $A(k)$ is to be determined; this term is included to accommodate the contribution from the homogeneous equation to the solution for $\hat{R}$. We choose $A(k)$ to be independent of ε. The asymptotic behaviours of $\hat{\psi}$ and $\hat{R}$, as $x \to \pm\infty$, are given in §4.3 as

$$\hat{\psi}(x; t, k, \varepsilon) \sim e^{-ikx} + be^{ikx}, \qquad \hat{R}(x, t; k, \varepsilon) \sim (b_t - 4ik^3 b)e^{ikx} + 4ik^3 e^{-ikx},$$
$$\text{as } x \to +\infty$$

$$\hat{\psi}(x; t, k, \varepsilon) \sim ae^{-ikx}, \qquad \hat{R}(x, t; k, \varepsilon) \sim (a_t + 4ik^3 a)e^{-ikx}, \qquad \text{as } x \to -\infty,$$

and to be consistent we must have

$$\hat{r}(x, t; k, \varepsilon) \sim \alpha^{\pm}e^{-ikx} + \beta^{\pm}e^{ikx} \qquad \text{as } x \to \pm\infty,$$

where the signs are ordered vertically. These asymptotic behaviours must satisfy equation (7.24), and so

$$\alpha^+ = \beta^- = 0, \qquad A(k) = 4ik^3, \qquad \text{for all } k,$$

which then give

$$a_t = \varepsilon\alpha^-, \qquad b_t - 8ik^3 b = \varepsilon\beta^+. \tag{7.25}$$

Note that these equations agree with the theory of §4.3 if we set $\varepsilon = 0$. The functions α^- and β^+ are obtained by solving

$$\hat{r}_{xx} + (k^2 - u)\hat{r} = q\hat{\psi}, \tag{7.26}$$

as we shall indicate below.

The asymptotic behaviour of the normalised discrete eigenfunction, ψ_n, is

$$\psi_n(x; t, \varepsilon) \sim c_n \exp(-\kappa_n x) \qquad \text{as } x \to +\infty, \tag{7.27}$$

and so equation (7.22) has a solution such that

$$R_n(x, t; \varepsilon) \sim \alpha_n \exp(-\kappa_n x) + \beta_n x \exp(-\kappa_n x) \qquad \text{as } x \to +\infty. \qquad (7.28)$$

The term in β_n is necessary since the asymptotic forcing term, $\exp(-\kappa_n x)$, in the differential equation, is also an asymptotic solution of the homogeneous equation for R_n. Upon the substitution of the behaviours (7.27) and (7.28) into equation (7.22) we obtain

$$2\kappa_n \beta_n = \varepsilon \omega_n c_n, \qquad (7.29)$$

since both u and q vanish at $+\infty$. These same behaviours and the definition of R yield

$$\alpha_n \exp(-\kappa_n x) + \beta_n x \exp(-\kappa_n x)$$
$$\sim c_{nt} \exp(-\kappa_n x) - c_n \kappa_{nt} x \exp(-\kappa_n x) - 4\kappa_n^3 c_n \exp(-\kappa_n x)$$

from which we obtain

$$\beta_n = -c_n \kappa_{nt} \qquad (7.30)$$

and

$$c_{nt} - 4\kappa_n^3 c_n = \alpha_n. \qquad (7.31)$$

Equations (7.29) and (7.30) are equivalent. In the absence of a perturbation term then $\varepsilon = 0$ and $\alpha_n = \beta_n = 0$, and so equation (7.31) recovers the result given in §4.3; equation (7.29) is identically satisfied.

The procedure for solving this perturbation problem can be easily described, although the details may prove laborious in specific cases. Let us suppose that we have a solution of the KdV equation (i.e. equation (7.18) with $\varepsilon = 0$), which we wish to perturb for small ε. This solution will therefore be the first term of an asymptotic expansion as $\varepsilon \to 0$, but it may also itself evolve on some suitable long scale. For this solution we find the corresponding eigenfunctions, ψ_n and $\hat{\psi}$, which are then used in equations (7.20), (7.22) and (7.26). The solutions of these equations enable $\kappa_n(t; \varepsilon)$, $b(k; t, \varepsilon)$ and $c_n(t; \varepsilon)$ to be determined to leading order in the perturbation. Upon reconstructing $u(x, t; \varepsilon)$ we shall obtain the leading order perturbation to the basic solution. This process may be continued, in principle, by successively iterating on the preceding solution.

7.2.1 Perturbation theory: an example

In order to illustrate how this iteration scheme is implemented in practice, we shall outline the construction of the solution for the case

$$q(x, t; \varepsilon) = \gamma(t) u(x, t), \qquad (7.32)$$

and with the basic solution taken to be a solitary wave. The solitary-wave solution is

$$u(x, t) = -2l^2 \operatorname{sech}^2(l\xi),$$

where $\xi = x - 4l^2 t$, for some positive constant l. We therefore express the leading term of the asymptotic solution as

$$u \sim u_0(x, t; \varepsilon) = -2l^2 \operatorname{sech}^2(l\xi) \tag{7.33}$$

where

$$\xi = x - x_0(t; \varepsilon), \quad \frac{dx_0}{dt} = 4l^2 + O(\varepsilon) \quad \text{and} \quad l = l(t; \varepsilon).$$

(This choice of solution allows for any 'slow' evolution on a suitably long timescale.) It is supposed that the initial value of l is given as $l(0; \varepsilon) = l_0$, which is not a function of ε.

The continuous eigenfunction associated with solution (7.33) is

$$\hat{\psi}_0(x; t, k, \varepsilon) = \left(\frac{k - il \tanh(l\xi)}{k - il} \right) \exp\{-ik(\xi + x_0)\},$$

so that

$$a(k; t, \varepsilon) \to \frac{(k + il)}{(k - il)} \quad \text{and} \quad b(k; t, \varepsilon) \to 0 \quad \text{as } \varepsilon \to 0.$$

This expression for the transmission coefficient, a, confirms that we have a discrete eigenvalue at $k = il$ ($l > 0$). Furthermore, we see that we require

$$a(k; 0, \varepsilon) = \frac{(k + il_0)}{(k - il_0)} \quad \text{and} \quad b(k; 0, \varepsilon) = 0,$$

to be consistent with the initial data.

The first approximation to $\hat{r}$ (see equation (7.24)) is obtained by solving equation (7.26) in the form

$$\hat{r}_{0xx} + (k^2 - u_0)\hat{r}_0 = \gamma u_0 \hat{\psi}_0,$$

where $\hat{r} \sim \hat{r}_0$ as $\varepsilon \to 0$, and

$$\hat{r}_0 \sim \begin{cases} \beta_0^+ e^{ikx} & \text{as } x \to +\infty \\ \alpha_0^- e^{-ikx} & \text{as } x \to -\infty. \end{cases}$$

The solution for $\hat{r}_0$ (obtained by the variation of parameters, for example) yields

$$\alpha_0^- = -\frac{2i\gamma l(3k^2 + l^2)}{3k(l + ik)^2}, \qquad \beta_0^+ = -\tfrac{2}{3}\gamma\pi \exp(2ikx_0) \operatorname{cosech}(\pi k/l).$$

(The details are rather complicated, even though it turns out that the complete solution need not be constructed in order to derive the asymptotic behaviours.) From equation (7.25) we therefore obtain

$$a_t = -\frac{2\varepsilon i \gamma l(3k^2 + l^2)}{3k(l + ik)^2} + o(\varepsilon) \tag{7.34}$$

and

$$b_t - 8ik^3 b = -\frac{2\varepsilon}{3}\gamma\pi \exp(2ikx_0)\operatorname{cosech}(\pi k/l) + o(\varepsilon) \qquad \text{as } \varepsilon \to 0. \tag{7.35}$$

The relevant solution for the discrete spectrum is far more straightforward. The normalised discrete eigenfunction associated with the single eigenvalue l is

$$(\tfrac{1}{2}l)^{1/2}\operatorname{sech}(l\xi);$$

see §4.5, example (i), and Q3.6. Thus we write

$$\psi_1 \sim \psi_{10} = (\tfrac{1}{2}l)^{1/2}\operatorname{sech}(l\xi) \qquad \text{as } \varepsilon \to 0,$$

and since

$$\omega_n = \int_{-\infty}^{\infty} \psi_n^2 q \, dx,$$

(see equations (7.20), (7.21)), we obtain

$$\omega_1 \sim -\gamma l^3 \int_{-\infty}^{\infty} \operatorname{sech}^4\{l(x - x_0)\} \, dx$$

$$= \tfrac{4}{3}\gamma l^2.$$

Thus the discrete eigenvalue evolves such that

$$l_t = \tfrac{2}{3}\varepsilon\gamma l + o(\varepsilon), \tag{7.36}$$

which, together with

$$c_1(t; \varepsilon) \sim (\tfrac{1}{2}l)^{1/2}, \tag{7.37}$$

gives us sufficient information to begin the reconstruction of the solution of the perturbed KdV equation. Notice that, to this order at least, it is not necessary to solve for R_n; the conditions given in equations (7.29) and (7.31) therefore *determine* the coefficients in equation (7.28) in terms of $l(t; \varepsilon)$.

The problem now has become one of obtaining the asymptotic solution, as $\varepsilon \to 0$, of the equations (7.34), (7.35) and (7.36); their solution constitutes the scattering data (although a is not used in the inverse scattering transform). We shall not pursue the details further, but a few comments must be made in conclusion. The discrete eigenvalue, from equation (7.36), becomes

$$l(t; \varepsilon) \sim l_0 \exp\left\{\tfrac{2}{3}\varepsilon \int_0^t \gamma(t') \, dt'\right\} \qquad \text{as } \varepsilon \to 0 \text{ at fixed } t,$$

since $l(0; \varepsilon) = l_0$. So if, for example, $\gamma(t) > 0$ for $t > 0$, we anticipate difficulties as $t \to \infty$: the asymptotic solution may not be uniformly valid. Indeed, if we write

$$a(k; t, \varepsilon) = \left(\frac{k + il}{k - il} \right)(1 + \varepsilon \hat{a}) \qquad (7.38)$$

then equations (7.34) and (7.36) show that

$$\hat{a}_t \sim \tfrac{2}{3} i \gamma l / k \qquad \text{as } \varepsilon \to 0, \qquad (7.39)$$

so that the expansion implied by equation (7.38) will certainly not be uniformly valid if there exists $\lim_{\varepsilon \to 0} \varepsilon \hat{a} \neq 0$. Furthermore, the behaviour given in (7.39) indicates that there is also a non-uniformity as $k \to 0$. It is therefore necessary to examine the whole solution described here for the case of small k. (We have throughout tacitly assumed that k is fixed as $\varepsilon \to 0$, and the solutions do turn out to be uniformly valid as $k \to \infty$.) These difficulties, and other points of interest, are discussed by, for example, Kaup & Newell (1978), Karpman & Maslov (1978), Knickerbocker & Newell (1980), Candler & Johnson (1981).

7.3 Numerical methods

This section is a brief introduction to numerical methods of solving nonlinear evolution equations which admit solitary-wave, soliton and such like solutions. Numerical calculation of solutions and their graphical presentation are informative and instructive. Indeed, it was the numerical experiments of Zabusky & Kruskal (1965) which initiated the development of the concept of the soliton (see §1.3) and the theory of the inverse scattering transform. Later, numerical calculations were used to indicate whether nonlinear wave solutions behave like solitons. In recent research the scattering transform and its analysis have been much more important than numerical methods. However, many nonlinear evolution equations do not have soliton solutions, and nonlinear equations with two or three spatial coordinates are little understood, so numerical results have still much to reveal.

After the development of numerical analysis, computers and graphics in the last twenty years, the calculation of soliton solutions has become easier. Nowadays such calculations provide suitable projects for undergraduate students. So here we shall discuss a few numerical methods suitable for some of the evolution equations with one space coordinate which we have met, and give references for further details. This will be an introductory manual to enable the reader to write programs and compute solutions for a project or research.

Nonlinear equations can be expressed in the form (5.23), namely

$$u_t = N(u), \tag{7.40}$$

where u may be a scalar or a vector field, and N is a nonlinear operator involving the spatial derivatives only. Also initial conditions and boundary conditions must be prescribed in order to determine uniquely the solution u (at least in principle). For computation, we must take finite boundaries, albeit distant ones; for a single spatial coordinate, x say, we may take them at $x = 0, 2\pi$ without loss of generality, by translation and scaling of the coordinate if necessary. Then the initial condition is the prescription of $u(x, 0)$ for $0 \leqslant x \leqslant 2\pi$. It is sometimes convenient to use periodic boundary conditions or to move the boundaries in order to track a soliton as it progresses.

We seek to approximate the exact solution $u(x, t)$ for $t > 0$, $0 \leqslant x \leqslant 2\pi$ by calculating the solution $\tilde{u}(x, t)$ of some numerical scheme. We require the error $\tilde{u}(x, t) - u(x, t)$ to be small, and ideally to converge uniformly to zero as a space step, h say, and a time step, k say, tend to zero. Little is known theoretically about the convergence of $\tilde{u}$ to u or about the stability of schemes for nonlinear equations, but much is known for equations linearised about some constant solution. Implicit schemes are often unconditionally stable. For linearised stability of an explicit scheme there is usually a condition giving an upper bound on the time step, the bound increasing as the space step does. Experience suggests that the stability criterion for the linearised scheme is a good guide to the corresponding criterion for the nonlinear scheme.

If exact conservation relations are known analytically for equation (7.40) then they give useful tests of the accuracy of a numerical solution. Also the numerical scheme may be devised to ensure that an important conserved quantity, mass or momentum, for example, is conserved to a high order of accuracy, with a truncation error which is either zero or small.

In any event, there are many good methods to solve an equation of the form (7.40), each with its own merits and its own advocates. We shall give the gist of a few good methods. To do this it helps to note that they belong essentially to one of two classes: *spectral methods* and *finite-difference methods*. With a spectral method the solution is approximated by some finite linear combination of a suitable set of functions, each one of which satisfies the boundary conditions. With a finite-difference method the derivatives in equation (7.40) are approximated by some differences to give a difference equation instead of a differential equation to be solved. After describing the methods we shall use them to solve the familiar nonlinear evolution equations (see Chap. 8).

7.3.1 Spectral methods

Choosing $\{\phi_p\}$ as any convenient complete set of functions which satisfy the spatial boundary conditions, we may expand any solution as

$$u(x,t) = \sum_{p=1}^{\infty} \hat{u}_p(t)\phi_p(x),$$

and hence seek to approximate the solution by

$$\tilde{u}(x,t) = \sum_{p=1}^{P} u_p(t)\phi_p(x) \qquad (7.41)$$

for a given 'large' integer P. There is some freedom in the choice of the basis $\{\phi_p\}$, but it is often useful to choose it as a set of trigonometric functions because the fast Fourier transform is an efficient and widely available algorithm. However, the basis could be chosen as a set of finite elements such as splines or other piecewise polynomials. In any event, the residual error may be defined by

$$r(x,t) = \tilde{u}_t(x,t) - N(\tilde{u}(x,t)) \qquad \text{for } 0 \leqslant x \leqslant 2\pi \qquad (7.42)$$

$$= \sum_{p=1}^{P} \frac{du_p}{dt}\phi_p(x) - N(\tilde{u}).$$

The coefficients u_p may be chosen in various ways in order to make r small, for example by Galerkin's method or by collocation. By Galerkin's method we impose the conditions that

$$\int_0^{2\pi} r(x,t)\phi_q(x)dx = 0 \qquad \text{for } q = 1,2,\ldots,P. \qquad (7.43)$$

Therefore

$$\sum_{p=1}^{P} a_{pq}\frac{du_p}{dt} = n_q(\tilde{u}), \qquad (7.44)$$

where

$$a_{pq} = \int_0^{2\pi} \phi_p(x)\phi_q(x)dx \qquad \text{and} \qquad n_q(\tilde{u}) = \int_0^{2\pi} N(\tilde{u})\phi_q(x)dx.$$

The method is more properly called *pseudospectral* than spectral if we use numerical integration rather than spectral analysis to evaluate n_q in terms of the coefficients u_p of the expansion (7.41). If the basis is orthogonal then we may normalise it so that $a_{pq} = \delta_{pq}$, the Kronecker delta, and equations

(7.44) become the system of ordinary differential equations,

$$\frac{du_q}{dt} = n_q(\tilde{u}) \qquad \text{for } q = 1, 2, \ldots, P, \tag{7.45}$$

where n_q is a function of the u_q. If the basis is not orthogonal then we may invert the $P \times P$ symmetric matrix with element a_{pq} in its pth row and qth column once and for all to derive a system of equations of the *form* (7.45). So we may integrate the system by a standard method, for example Runge–Kutta, to find the coefficients $u_p(t)$ for $t > 0$ and $p = 1, 2, \ldots, P$, substitute into the approximate solution (7.41) and hence evaluate $\tilde{u}(x, t)$.

7.3.2 Finite-difference methods

In a finite-difference method we specify the solution $u(x, t)$ numerically only at the discrete points $x = mh$ and $t = nk$ for $m = 0, 1, \ldots, M$ and $n = 0, 1, \ldots$, where $h = 2\pi/M$, say, is the space step and k the time step, and then devise a scheme to calculate u_m^n as an approximation to $u(mh, nk)$ for small h and k. Thus, using various truncated Taylor expansions to approximate the derivatives of u, we may approximate equation (7.40) by a difference equation. Some commonly used difference equations are of the forms

$$u_m^{n+1} = u_m^n + kN_h(u_m^n) \qquad \text{(forward-Euler scheme),} \tag{7.46a}$$

$$u_m^{n+1} - u_m^n = kN_h(u_m^{n+1}) \qquad \text{(backward-Euler scheme),} \tag{7.46b}$$

$$u_m^{n+1} = u_m^{n-1} + 2kN_h(u_m^n) \qquad \text{(leap-frog scheme),} \tag{7.46c}$$

and

$$u_m^{n+1} - u_m^n = \tfrac{1}{2}k\{N_h(u_m^n) + N_h(u_m^{n+1})\} \quad \text{(Crank–Nicolson scheme),} \tag{7.46d}$$

where $N_h(u_m^n)$ is some suitable approximation to $N(u)$ at $x = mh, t = nk$ which may involve $u_m^n, u_{m-1}^n, u_{m+1}^n, \ldots$ in order to represent the spatial derivatives of u. The forward-Euler and leap-frog schemes are explicit, giving the solution at the $(n + 1)$th time step explicitly in terms of the solution at earlier time steps, but the backward-Euler and Crank–Nicolson schemes are implicit. Also equation (7.46a) is used for odd values of $m + n$ and (7.46b) for even values in the hopscotch scheme. Sometimes u_m^{n+1} is replaced by $\tfrac{1}{2}(u_m^{n+1} + u_{m-1}^{n+1})$ in order that equation (7.46b) becomes linear in the 'unknowns' u_m^{n+1} for $m = 0, 1, \ldots, M$ and hence easily rendered explicit.

If the error for a scheme is

$$u_m^n - u(mh, nk) = O(h^\sigma + k^\tau) \qquad \text{as } h, k \to 0$$

then the scheme is said to be of order σ in space and τ in time. The forward-

and backward-Euler schemes are of first, and the leap-frog and Crank–Nicolson schemes of second, order in time. The Crank–Nicolson scheme is, moreover, unconditionally stable for the usual simple linear discrete forms of N_h.

An implicit scheme requires all the difference equations at one time step to be solved in order to advance to the next step. There are various algebraic methods to do this efficiently. The computational time lost in this way may, however, be offset by the implicit scheme's being more stable than an explicit one, so that a longer time step may be used.

7.3.3 *Long-wave equations*

For our first application of numerical methods we take the KdV equation. Zabusky & Kruskal (1965) originally solved it by using a leap-frog scheme. They approximated the equation,

$$u_t + uu_x + u_{xxx} = 0, \tag{7.47}$$

by

$$u_m^{n+1} = u_m^{n-1} - \tfrac{1}{3}kh^{-1}(u_{m+1}^n + u_m^n + u_{m-1}^n)(u_{m+1}^n - u_{m-1}^n)$$
$$- kh^{-3}(u_{m+2}^n - 2u_{m+1}^n + 2u_{m-1}^n - u_{m-2}^n). \tag{7.48}$$

This scheme conserves 'mass', so that $\sum_{m=0}^{M-1} u_m^n$ is independent of n. Their choice of the average $\tfrac{1}{3}(u_{m+1}^n + u_m^n + u_{m-1}^n)$ to approximate u in the nonlinear term uu_x was made to conserve the 'energy' to second order, giving

$$\frac{1}{2}\sum_{m=0}^{M-1}(u_m^{n+1})^2 - \frac{1}{2}\sum_{m=0}^{M-1}(u_m^{n-1})^2 = O(k^3) \qquad \text{as } k \to 0$$

if u is periodic (or vanishes near the ends of the interval of integration). Indeed, the whole scheme is second-order in time. The linearised stability condition of the scheme is

$$k \leqslant h^3/\{4 + h^2|U|\}, \tag{7.49}$$

where U is the constant solution about which there is linearisation (Greig & Morris 1976), and which may be taken as the maximum magnitude of u. Note that $k = O(h^3)$ as $h \to 0$, so that small time steps and therefore quite long computation times are required. Sanz-Serna (1982) has discussed the scheme critically and showed how to modify it to conserve 'energy' exactly.

Greig & Morris (1976) used the hopscotch scheme for (7.47),

$$u_m^{n+1} + \frac{k}{2h}\theta_m^{n+1}\left\{ H_x(\tfrac{1}{2}(u_m^{n+1})^2) + \frac{1}{h^2}H_x\delta_x^2 u_m^{n+1} \right\}$$
$$= u_m^n - \frac{k}{2h}\theta_m^n\left\{ H_x(\tfrac{1}{2}(u_m^n)^2) + \frac{1}{h^2}H_x\delta_x^2 u_m^n \right\}, \tag{7.50}$$

where the hopscotch switch is defined as

$$\theta_m^n = \begin{cases} 1 & \text{if } m+n \text{ is even} \\ 0 & \text{if } m+n \text{ is odd} \end{cases}$$

and the difference operators by

$$\delta_x u_m^n = u_{m+1/2}^n - u_{m-1/2}^n \qquad \text{and} \qquad H_x u_m^n = u_{m+1}^n - u_{m-1}^n$$

for all m, n. This scheme conserves energy to second order in time, and is stable if

$$k \leqslant h^3/|2 - h^2 U|, \tag{7.51}$$

a condition less stringent than (7.49), but nonetheless one for which $k = O(h^3)$ as $h \to 0$ and therefore one which requires quite long computation times.

Fornberg & Whitham (1978) used a pseudospectral method, taking a finite Fourier transform F of $\tilde{u}$ with respect to x, where the approximate solution $\tilde{u}$ of the KdV equation (7.47) is defined at $2P$ points of the interval $0 \leqslant x < 2\pi$ and has period 2π. Thus they expanded

$$\tilde{u}(qh, t) = F^{-1} u_p$$

$$= \frac{1}{(2P)^{1/2}} \sum_{p=-P}^{P} u_p(t) e^{\pi i pq/P} \qquad \text{for } q = 0, 1, \ldots, 2P - 1, \tag{7.52}$$

where only half of the contributions at $p = \pm P$ are included in the sum, $h = \pi/P$, and

$$u_p(t) = F\tilde{u}$$

$$= \frac{1}{(2P)^{1/2}} \sum_{q=0}^{2P-1} \tilde{u}(qh, t) e^{-\pi i pq/P}. \tag{7.53}$$

The fast Fourier transform is an efficient method to evaluate these series when P is a power of two. Then u_x is approximated by $F^{-1}(ipF\tilde{u})$ and u_{xxx} by $F^{-1}(-ip^3 F\tilde{u})$. With a leap-frog scheme, this gives

$$\tilde{u}(x, (n+1)k) - \tilde{u}(x, (n-1)k) + 2ik\tilde{u}F^{-1}(pF\tilde{u}) - 2ikF^{-1}(p^3 F\tilde{u}) = 0,$$

where $\tilde{u}$ and the inverse transforms are evaluated at $x = qh$ and $t = nk$. However, Fornberg & Whitham modified the last term, taking instead

$$\tilde{u}(x, (n+1)k) - \tilde{u}(x, (n-1)k) + 2ik\tilde{u}F^{-1}(pF\tilde{u}) - 2iF^{-1}\{\sin(p^3 k)F\tilde{u}\} = 0. \tag{7.54}$$

These two schemes are equivalent for small time steps, because $\sin(p^3 k) = p^3 k + O(k^3)$ as $k \to 0$, but the modified scheme (7.54) represents the *linearised* KdV equation, $u_t + u_{xxx} = 0$, exactly and therefore improves the accuracy for components of high wavenumbers p. The scheme requires three fast

Fourier transforms per time step, and its linearised stability condition is that

$$k \leqslant 3h^3/2\pi^2. \tag{7.55}$$

A comparative study (Taha & Ablowitz 1984b) shows the scheme to be faster than the finite-difference schemes.

The method of Fornberg & Whitham is also suitable for other 'long-wave' equations of the form

$$u_t + u^j u_x + \int_{-\infty}^{\infty} K(x - \xi) u_\xi(\xi, t) \mathrm{d}\xi = 0,$$

where j is a positive integer and the kernel K is determined by the dispersion relation of the linearised waves (cf. Q1.14, Q1.15).

When Peregrine (1966) proposed the regularised long-wave equation,

$$u_t + u_x + u u_x - u_{xxt} = 0,$$

to model an undular bore, he used the numerical scheme,

$$\left(1 - \frac{\delta_x^2}{h^2}\right)(u_m^{n+1} - u_m^n) + (1 + u_m^n)\frac{k}{2h} H_x \tfrac{1}{2}(u_m^{n+1} + u_m^n) = 0$$

to integrate the equation. This scheme is only of first order accuracy in time and second order in space, but is always linearly stable (Eilbeck & McGuire 1975, p. 46).

Eilbeck & McGuire (1975) proposed and used the scheme

$$\left(1 - \frac{\delta_x^2}{h^2}\right)(u_m^{n+1} - u_m^{n-1}) + \frac{k}{h}(1 + u_m^n)H_x u_m^n = 0,$$

which is of second order in both space and time. This seems to be more efficient, although it is a three-level scheme in time (involving $n + 1, n$ and $n - 1$) and requires an equation with a tridiagonal matrix of coefficients to be solved. The linearised stability condition is

$$k^2 \leqslant (4 + h^2)/(1 + U)^2,$$

and so is always satisfied for the small space and time steps required for accuracy in practice. These and other finite-difference schemes are critically examined by Bona, Pritchard & Scott (1985). Ben-yu & Manoranjan (1985) have solved the RLW equation by a pseudospectral method.

7.3.4 Nonlinear Klein–Gordon equations

Perring & Skyrme (1962) originally solved the sine–Gordon equation by using a leap-frog scheme. They approximated the equation

$$u_{xx} - u_{tt} = \sin u, \tag{7.56}$$

by using the finite-difference scheme,

$$u_m^{n+1} = -u_m^{n-1} + \frac{k^2}{h^2}(u_{m+1}^n + u_{m-1}^n) + 2\left(1 - \frac{k^2}{h^2}\right)u_m^n - k^2 \sin u_m^n. \quad (7.57)$$

The theory of linearised stability shows that this scheme is stable for $k < h$, and in practice it is very stable for $k < 0.95h$.

The sine–Gordon is the most celebrated of the nonlinear Klein–Gordon equations of the form,

$$u_{xx} - u_{tt} = F(u). \quad (7.58)$$

They all require similar numerical schemes, so a leap-frog scheme is again suitable. Ablowitz, Kruskal & Ladik (1979) improved the stability of the scheme of Perring & Skyrme, using the space average of $\frac{1}{2}(u_{m+1}^n + u_{m-1}^n)$ instead of u_m^n in $F(u)$, and taking $k = h$, for three examples. This gives

$$u_m^{n+1} = -u_m^{n-1} + u_{m+1}^n + u_{m-1}^n - h^2 F\{\tfrac{1}{2}(u_{m+1}^n + u_{m-1}^n)\}. \quad (7.59)$$

This allows the calculations to be performed with only even or only odd values of $m + n$, and so halves the computation time.

7.3.5 The nonlinear Schrödinger equation

Taha & Ablowitz (1984a) give details of eight schemes to solve the NLS equation, schemes spectral and finite-difference, implicit and explicit. They compared the computing times taken by the schemes to solve various initial-value problems. The pseudospectral method of Fornberg & Whitham (1978) or a split-step Fourier method was fastest for each problem.

Further reading

7.1 The Painlevé property is discussed by Ablowitz & Segur (1981, §3.7).

7.2 Further details of perturbation theory can be found in Ablowitz & Segur (1981, §3.8); Newell (1985, §3g).

7.3 Dodd, Eilbeck, Gibbon & Morris (1982, Chap. 10), have reviewed many numerical methods to solve nonlinear evolution equations, and show many of their solutions.

Exercises

(The reader is advised to recapitulate the problems Q1.13, Q2.9 and Q2.10.)

Q7.1 *Movable critical points.* Find the general solutions of these equations and hence decide if the solutions contain critical points, and if the points are movable:

(i) $2(z-1)\dfrac{dw}{dz} - w = 0;$

(ii) $2\dfrac{dw}{dz}\dfrac{d^3w}{dz^3} = 3\left(\dfrac{d^2w}{dz^2}\right)^2;$

(iii) $\dfrac{dw}{dz}\dfrac{d^3w}{dz^3} = 2\left(\dfrac{d^2w}{dz^2}\right)^2.$

Q7.2 *Singular-point analysis.* For each of the equations in Q7.1 assume that

$$w(z) \sim A(z - z_0)^n \qquad \text{as } z \to z_0,$$

(or, if this fails, $w(z) \sim A \log(z - z_0)$ as $z \to z_0$), where A, z_0 and n are constants. Hence decide if the solution has a critical point and if it is movable.

Now apply the same procedure to the P-I equation,

$$\frac{d^2w}{dz^2} = 6w^2 + z.$$

Q7.3 *Transformation to P-form.* Show that these equations reduce to Painlevé equations under the transformations given:

(i) $\dfrac{d^2\phi}{dz^2} = 6(z + \phi)^2 - \phi$ with $\phi(z) = \tfrac{1}{12} - z + w(z - \tfrac{1}{24});$

(ii) $9z^2\dfrac{d^2\phi}{dz^2} = (z - 2)\phi + 2\phi^3$ with $\phi(z) = z^{1/3}w(z^{1/3}).$

Q7.4 *Burgers equation.* Show that the travelling-wave solution (see Q2.1(i)) and the similarity solution (see Q2.8) of the Burgers equation

$$u_t + uu_x = vu_{xx},$$

where v is a constant, satisfy, after one integration, appropriate generalised Riccati equations (see equation (7.2)).

Q7.5 *Sine–Gordon equation.* Seek a similarity solution of the sine–Gordon equation,

$$u_{xt} = \sin u,$$

in the form $u(x,t) = F(xt^n)$, for suitable n, and hence show that the transformation

$$w(z) = \exp\{iF(z)\}$$

yields a Painlevé equation for w.

Q7.6 *Nonlinear Schrödinger equation.* Seek a solution of the NLS equation

$$iu_t + u_{xx} + vu|u|^2 = 0,$$

where v is a constant, in the form

$$u(x,t) = F(x)e^{i\lambda t},$$

where λ is a real constant. Obtain the equation for $F(x)$ and, assuming that F is real, use the singular point analysis of Q7.2 to examine the local nature of the solution.

Q7.7 *A nonlinear Klein–Gordon equation.* Seek a similarity solution of the equation

$$u_{tt} - u_{xx} = u^N,$$

where $N(>1)$ is an integer, in the form

$$u(x, t) = t^m F(xt^n)$$

for suitable values of m and n. Obtain the equation for $F(z)$, and for the case $N = 3$ use the singular point analysis of Q7.2 to examine the local nature of the solution.

Q7.8 *Sinh–Gordon equation.* Use the approach adopted in Q7.5 to discuss the equation

$$u_{xt} = \sinh u.$$

Q7.9 *Two-dimensional KdV equation.* Use the similarity form

$$u(x, t, y) = t^n F(xt^p + \lambda y^2 t^q),$$

for suitable n, p, q and λ, and construct the equation for F if u satisfies

$$(u_t - 6uu_x + u_{xxx})_x + 3u_{yy} = 0;$$

see Q5.14 and §6.2.3(b). Integrate your equation for $F(z)$ once and suppose that $F, F', \ldots \to 0$ somewhere. Now multiply by F, integrate again, and show that v satisfies the P-II equation with $F = v^2$.

8

Epilogue

In this final chapter we shall first use the numerical methods described in §7.3, discussing a few case studies of solitons and their interaction; this will also illustrate some of the analytic results of earlier chapters. Then we shall list briefly the major applications of the chief nonlinear evolution equations.

8.1 Some numerical solutions of nonlinear evolution equations

We begin by taking a final look at the two-soliton interaction for the KdV equation. Hirota's method (see §5.3.2 and Q4.3) for the KdV equation in the form

$$u_t - 6uu_x + u_{xxx} = 0 \qquad (8.1)$$

gives the two-soliton solution

$$u(x,t) = -2\frac{k_1^2 E_1 + k_2^2 E_2 + 2(k_2 - k_1)^2 E_1 E_2 + A_2(k_2^2 E_1 + k_1^2 E_2)E_1 E_2}{(1 + E_1 + E_2 + A_2 E_1 E_2)^2}$$

$$(8.2)$$

where $E_i = \exp(\theta_i)$, $\theta_i = k_i x - k_i^3 t + \alpha_i$ $(i = 1, 2)$ and $A_2 = (k_2 - k_1)^2/(k_1 + k_2)^2$. This solution for $k_1 = 1$, $k_2 = \sqrt{2}$, $\alpha_1 = \alpha_2 = 0$ is pictured in Fig. 8.1 in a frame with respect to which the slower (shorter) soliton is at rest. The re-assumption of the identities of the two solitary waves after their interaction is vividly shown, and so are the phase shifts of the two waves. It can be seen that the taller wave approaches the smaller wave and then, at a short distance, the waves exchange rôles: the smaller wave grows taller, and the taller becomes smaller, before they separate. This solution also illustrates the point of Q4.5: the two-soliton solution may have either one or two local maxima. In fact (Lax, 1968), if $k_2^2/k_1^2 > \frac{1}{2}(3 + \sqrt{5})$ then the taller wave first absorbs, then re-emits the shorter wave, i.e. the fast wave appears to 'pass through' the slow wave, as shown in Fig. 4.3.

Fig. 8.1 Perspective view of the two-soliton solution, (8.2), with $k_1 = 1$ and $k_2 = \sqrt{2}$, where $X = x - t$.

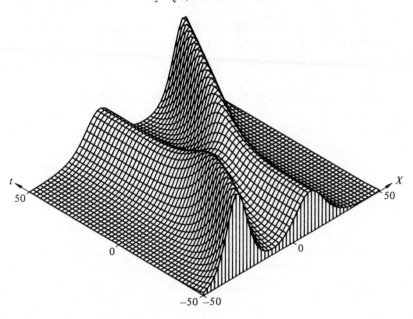

Fig. 8.2 Perspective view of the modulus of a solution of an NLS equation, representing the interaction of two solitons. The plot is of $|u(x, t)|$ versus x and t, where u satisfies equation (8.3) and

$$u(x, 0) = \sqrt{2}\{e^{i(x-5)/2} \operatorname{sech}(x - 5) + e^{-i(x-15)/2} \operatorname{sech}(x - 15)\}.$$

The maximum value of $|u|$ found on the grid is 2.8313 at $x = 10$, $t = 4.2$.

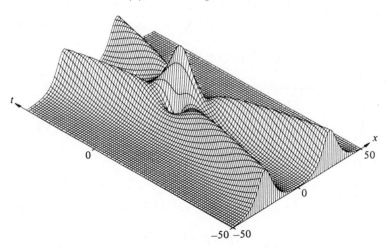

A solution of the NLS equation

$$iu_t + u_{xx} + u|u|^2 = 0 \qquad (8.3)$$

is shown in Fig. 8.2. Recall that a solitary-wave solution, given in Q2.3, is a complex function; it describes a carrier wave with an amplitude modulation which behaves rather like a KdV solitary wave. The perspective view given in Fig. 8.2 shows a solution which is close to that of the interaction of two solitons. The solution has been found by numerical integration of the equation, using a pseudospectral method, for the given initial values of $u(x,0)$, which are the *linear* superposition of two fairly distinct solitons.

Fig. 8.3 The interaction of two kinks. A solution $\phi(x,t)$ of the sine–Gordon equation (8.4) is plotted versus x for (a) $t = 0$, (b) $t = 5$, (c) $t = 10$, (d) $t = 15$ and (e) $t = 20$. (The solution is close to the analytic solution

$$\varphi(x,t) = 4\arctan\{\lambda \sinh(x/(1-\lambda^2)^{1/2})/\cosh(\lambda t/(1-\lambda^2)^{1/2})\}$$

of Perring & Skyrme (1962) with $\lambda = \frac{1}{2}$ after a time shift of 10.) The kinks proceed with velocities $\pm\frac{1}{2}$ except when they interact.

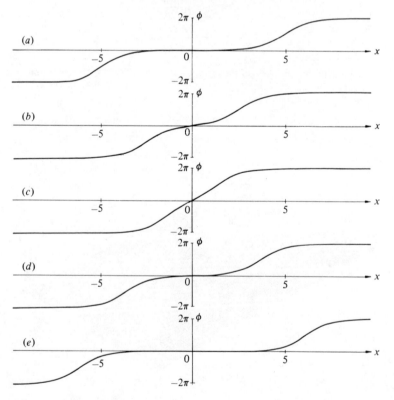

Next we shall show a few examples of solutions of the sine–Gordon equation in the form

$$\phi_{xx} - \phi_{tt} = \sin\phi \qquad (8.4)$$

(i.e. in laboratory coordinates), which have been found by use of a leap-frog scheme. But first note that the simplest soliton is a kink (see Q2.19), i.e. a solution (equivalent to (5.71)) for which

$$\phi(x, t) = 4\arctan\left[\exp\left\{(x - \lambda t)/(1 - \lambda^2)^{1/2}\right\}\right] \qquad (8.5)$$

and ϕ increases monotonically from zero to 2π as x increases from $-\infty$ to ∞, when $-1 < \lambda < 1$. There also exist antikinks for which ϕ decreases from 2π to zero as x increases from $-\infty$ to ∞. There are similar kinks

Fig. 8.4 The interaction of a kink and an antikink. A solution $\phi(x, t)$ of equation (8.4) is plotted for (a) $t = 0$, (b) $t = 7.5$, (c) $t = 15$, (d) $t = 22.5$ and (e) $t = 30$. The kink advances with velocity 0.6 and the antikink with velocity -0.5 except when they interact, and the solution vanishes instantaneously between $t = 0$ and $t = 7.5$. (This solution, after a Lorentz transformation, is essentially the analytic solution given in Q 2.22.)

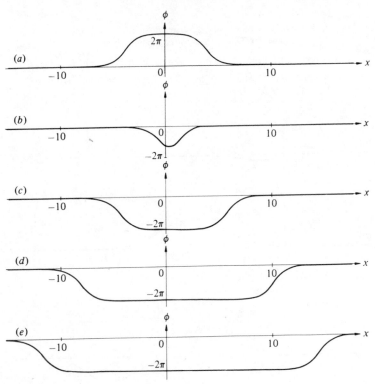

Fig. 8.5 A stationary breather. This is a sketch of the analytic solution of the sine–Gordon equation (8.4), given in Q2.20 with $t_0 = x_0 = 0$, $\lambda = 0.2$, for various instants.

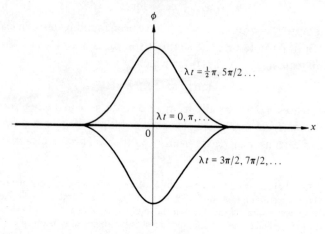

Fig. 8.6 A moving breather: $\phi(x,t)$ versus x for (a) $t = 0.1$, (b) $t = 10$ and (c) $t = 20$. This is close to the analytic solution,

$$\phi(x,t) = 4 \arctan\left[(1 - \lambda^2)^{1/2} \sin\{\gamma\lambda(t - Vx)\}/\lambda \cosh\{\gamma(1 - \lambda^2)^{1/2}(x - Vt)\}\right]$$

of equation (8.4), where $\gamma = 1/(1 - V^2)^{1/2}$, for $V = 0.5$ and $\lambda = 0.2$.

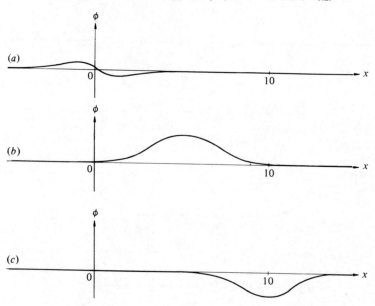

and antikinks for which an integral multiple of 2π is added to ϕ. With this background we may illustrate Q6.23 and interpret Fig. 8.3, which shows the interaction of two kinks, and Fig. 8.4 which shows the interaction of a kink and an antikink. The character of the kinks and antikinks as topological solitons can be seen clearly as well as their soliton properties before and after their brief periods of strong interaction.

A stationary breather of the sine–Gordon equation (8.4) (see Q2.20) is shown in Fig. 8.5 (equivalent to the breather of Q6.21) and a moving breather (see Q2.21) in Fig. 8.6. The two are related by a Lorentz transformation of the frames of reference. With this in mind we may interpret the interaction of a kink and a moving breather shown in Fig. 8.7. Although the solution is shown at only a few instants, the soliton property of the strong interaction is clearly visible.

Finally we present two solutions associated with waves which propagate in two spatial dimensions. The first (Fig. 8.8) shows a rational solution

Fig. 8.7 The interaction of a kink with velocity $\frac{1}{2}$ and a breather with velocity $-\frac{1}{2}$ and frequency $\lambda = 0.2$. This shows $\phi(x, t)$ versus x for (a) $t = 0$, (b) $t = 7.5$, (c) $t = 15$, (d) $t = 22.5$ and (e) $t = 30$.

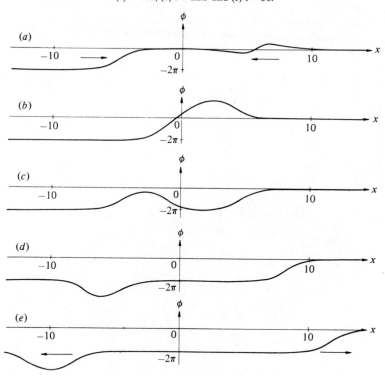

of the two-dimensional KdV equation

$$(u_t - 6uu_x + u_{xxx})_x - 3u_{yy} = 0$$

(see Q2.24). This solution is sometimes called the Zakharov–Manakov (ZM) rational soliton. It propagates with unchanging form, and may interact with other solutions of the equation to give various types of N-soliton solution; see Freeman (1980). The second (Fig. 8.9) depicts a resonant interaction (see Q8.3) of the other 2D KdV equation

$$(u_t - 6uu_x + u_{xxx})_x + 3u_{yy} = 0.$$

The three 'arms' of this solution extend to infinity and the whole structure constitutes a steadily progressing wave. This wave may interact with other solitons and, indeed, with other resonant solutions (Anker & Freeman, 1978b; Freeman, 1980).

8.2 Applications of nonlinear evolution equations

We have related how Korteweg and de Vries discovered their equation as a model of small amplitude waves on shallow water, and described

Fig. 8.8 Perspective view of a rational solution of the 2D KdV equation $(u_t - 6uu_x + u_{xxx})_x - 3u_{yy} = 0$ with $p = 1$ (see Q2.24), on plotting $u(X, y)$ where $X = x + 1 - 3t$.

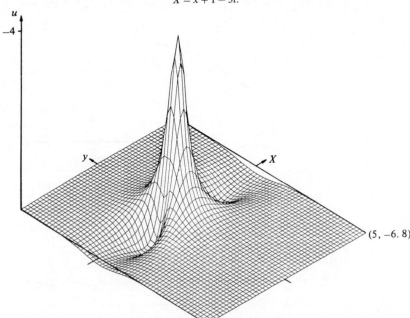

many other applications of the KdV equation in §§1.2 and 1.4. The applications are ubiquitous because the KdV equation is a canonical equation to describe the amplitude of weakly nonlinear long waves in a moving medium. This combination of the leading effects of weak non-linearity and weak dispersion of long waves is emphasised in Q1.14 as well as §1.4.

For similar reasons the Benjamin–Ono equation describes some weakly nonlinear long internal gravity waves. The weak nonlinearity is of the same form as for water waves but the dispersion relation for long linear waves differs, and so the KdV equation does not arise (see Q1.15).

Fig. 8.9 Perspective view of a resonant interaction with $m_1 = m_2 = 1$, $n_1 = 1$, $n_2 = 2$ (see Q8.3). The amplitudes of the three 'arms' far from the intersection are -2, -8, -18.

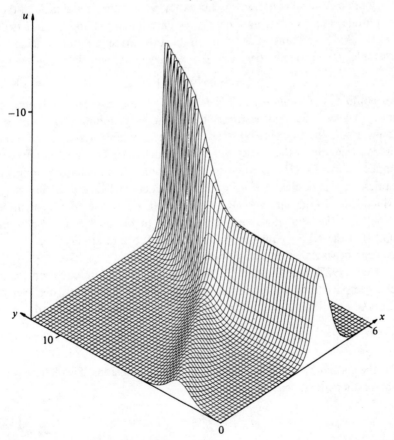

If, however, the leading approximation to the weak nonlinearity is of higher order than quadratic then a modified KdV equation may arise. Thus a modified KdV equation governs the modulation of the amplitude A of many weakly nonlinear long waves for which the coefficient of AA_x happens to be zero. Early uses of the modified KdV equations were to model phonons in an anharmonic (i.e. nonlinear) lattice (Zabusky, 1967) and an Alfvén wave in a collisionless plasma (Kakutani & Ono, 1969). However, the quadratic modified KdV equation has attracted attention mostly because of its close mathematical relation to the KdV equation (see §§4.2, 5.4.2).

The nonlinear Schrödinger equations similarly have diverse applications because they are canonical equations. There are essentially only two different equations after scaling transformations, namely

$$iA_t + A_{xx} \pm A|A|^2 = 0. \tag{8.6a, b}$$

They are canonical equations because they govern the modulation of the amplitude A of a weakly nonlinear wave packet in a moving medium (see Q8.1). (An important generalisation of this argument for a slightly unstable wave gives the *Ginzburg–Landau equation*, namely

$$A_\tau + aA_{\xi\xi} = bA + cA|A|^2, \tag{8.7}$$

for complex constants a and b determined by the dispersion relation of linear waves, and c determined by the weakly nonlinear interaction (Stewartson & Stuart, 1971). However, the Ginzburg–Landau equation arose originally in the theory of superconductivity.) The most important applications of the NLS equation are to some phenomena of nonlinear optics. In this context we write of the self-focussing of the carrier wave which leads to the *envelope-soliton* solution (Q2.3) of the NLS$^+$ equation (8.6a) and the self-defocussing which leads to the *dark-soliton* solution (Q2.4) of the NLS$^-$ equation (8.6b). Other important applications are to packets of water waves and plasma waves.

Klein (1927) and Gordon (1926) derived a relativistic equation for a charged particle in an electromagnetic field, using the recently discovered ideas of quantum theory. Their *Klein–Gordon equation* reduces to

$$\frac{1}{c^2}\frac{\partial^2\psi}{\partial t^2} - \nabla^2\psi + \left(\frac{mc}{\hbar}\right)^2\psi = 0 \tag{8.8}$$

for the special case of a free particle in three dimensions. This later led to the mathematical generalisation

$$\frac{1}{c^2}\frac{\partial^2\psi}{\partial t^2} - \nabla^2\psi + V'(\psi) = 0 \tag{8.9}$$

for some differentiable potential function V; this is called a *nonlinear Klein–Gordon equation* when V' is a nonlinear function. Of course, the equation is invariant under the Lorentz transformation (Q2.21). In particular, if $V'(\psi) = \sin \psi$ and we restrict the equation to one spatial dimension, then we get the sine–Gordon equation

$$\frac{1}{c^2} \psi_{tt} - \psi_{xx} + \sin \psi = 0. \tag{8.10}$$

This is in 'laboratory' coordinates. With 'characteristic' coordinates $u = \frac{1}{2}(x - ct)$ and $v = \frac{1}{2}(x + ct)$ it becomes

$$\psi_{uv} = \sin \psi, \tag{8.11}$$

and clearly an initial-value problem with one form of the equation is not an initial-value problem with the other.

As we mentioned in §5.4, the sine–Gordon equation first arose last century in the theory of differential geometry. It arises as a direct model of physical phenomena also, and as another evolution equation for the amplitude of various slowly varying waves. Frenkel & Kontorova (1939) found that the sine–Gordon equation governs the propagation of a dislocation in a crystal whose periodicity is represented by $\sin \psi$; Perring & Skyrme (1962) posed it as a tentative model of an elementary particle; and Coleman (1975) showed that it is an equivalent form of the Thirring model. Again, it governs the propagation of magnetic flux in a long Josephson-junction transmission line, where $\sin \psi$ is the Josephson current across an insulator between two superconductors and the voltage is proportional to ψ (cf. Bhatnagar, 1979, p.39); this offers good examples of the experimental observation of soliton interactions. Gibbon, James & Moroz (1979) showed that the sine–Gordon equation governs the modulation of a weakly unstable baroclinic wave packet in a two-layer fluid, and thence similar wave packets in a moving medium.

Kadomtsev & Petviashvili (1970) showed that the equations

$$(u_t - 6uu_x + u_{xxx})_x \pm 3u_{yy} = 0$$

govern slowly varying waves in dispersive media. The equation with the upper sign $(+)$ arises in the study of plasmas, and also in the modulation of long weakly nonlinear water waves which propagate nearly in one-dimension (i.e. nearly in a vertical plane). The equation then has solutions which describe the oblique interaction of solitons. The equation with the lower sign $(-)$ arises in acoustics. This equation has rational solutions, whereas the equation with the upper sign does not (see Q2.24).

Research into the physical, earth and life sciences has led to the study of hundreds more nonlinear evolution equations. Of these equations only a few score are known to have soliton solutions. We have had space enough to treat only a few of the equations and to cover a few more, briefly, in the exercises. The recommendations for further reading, however, will help readers who wish to follow up other equations and applications.

Yet one other field of application deserves special mention. The superficial similarity between the properties of solitons and of elementary particles is striking. Solitons may propagate without change of form. A soliton may be regarded as a local confinement of the energy of the wave field. When two solitons collide, each may come away with the same character as it had before the collision. When a soliton meets an antisoliton, both may be annihilated. Elementary particles share these properties. So, if an appropriate system of nonlinear field equations admits soliton solutions then these solitons may represent elementary particles and have properties which may be confirmed by observations of particles. Quantum field theories are deep and complicated, so research into their solitary-wave solutions, although intense now, is at a much more primitive stage than that into the simple soliton solutions of the KdV, NLS and sine–Gordon equations. In particular, the Yang–Mills field equations seem to be a fruitful model unifying electromagnetic and weak forces. They admit solutions localised in space which represent very heavy elementary particles. They also have solutions, called *instantons*, localised in time as well as space; these are interpreted as quantum-mechanical transitions between different states of a particle. In further reading on solitons as a topic in the theory of elementary particles you may note that particle physicists define a soliton merely as a localised solution of permanent form, i.e. they omit our defining property (iii) at the end of §1.3. Their models are too complicated to enable an inverse scattering transform to be devised at present, although some topological properties of the solutions are useful. A popular introduction to recent research into the field has been written by Rebbi (1979), opening up fascinating vistas.

Further reading

8.2 Bhatnagar (1979, Chap. 2) lists may references about applications of nonlinear wave equations. Ablowitz & Segur (1981, Chap. 4), describes diverse applications of many equations in some detail. Craik (1985) derives several canonical amplitude equations governing the weakly nonlinear interactions of waves, especially those in fluids and plasmas.

Lee (1981), Rajaraman (1982) and Vainshtein, Zakharov, Novikov & Shifman (1982) treat solitons and instantons in the theory of particle physics.

Exercises

Q8.1 *The modulation of a wave packet.* A real linear partial differential equation is given in the operational form

$$u_t + \mathrm{i}f(-\mathrm{i}\partial/\partial x)u = 0,$$

where f is an odd function. Show that wave solutions $u(x,t) = \mathscr{R}\{\mathrm{e}^{\mathrm{i}(kx-\omega t)}\}$ have the dispersion relation $\omega = f(k)$.

Consider a wave packet described by $u(x,t) = \mathscr{R}\{A(x,t)\,U(x,t)\}$, where the amplitude A is a slowly varying complex function, and the 'carrier wave' $U = \mathrm{e}^{\mathrm{i}(Kx-\Omega t)}$ has frequency $\Omega = f(K)$ for a given wave number $K \neq 0$. Deduce that

$$\mathrm{i}\{A_t + f'(K)A_x\} + \tfrac{1}{2}f''(K)A_{xx} = 0$$

approximately, if f has a Taylor series about K.

You are given further that the addition of some nonlinear terms to the equation for u results, at leading order for small amplitudes A, in the addition of a term proportional to $|A|^2A$ to the modulation equation for A, so weakly nonlinear wave packets are governed by the equation,

$$\mathrm{i}\{A_t + f'(K)A_x\} + \tfrac{1}{2}f''(K)A_{xx} + l|A|^2A = 0.$$

Remove the term in A_x by a Galilean transformation (noting that $f'(K)$ is the group velocity of the linear wave packet) and rescale x and t to show that this equation is equivalent to an NLS equation.

[The term in $|A|^2A$ arises generally because it is the leading approximation to the self-interaction of a weakly nonlinear wave in a homogeneous medium, homogeneity implying that wave propagation is invariant under translation.]

Q8.2 *A simple experimental model for the sine–Gordon equation.* Obtain a length (100–150 cm, say) of rubber band about 6 mm wide and 2 mm thick. Insert carefully a set of similar pins with large heads into one side of the strip, placing them centrally and uniformly about 3 mm apart. Clamp each end of the strip so that the pins hang downwards as a row of simple pendulums, as illustrated in Fig. 8.10.

Fig. 8.10 A sketch of a short stretch of the rubber band described in Q8.2.

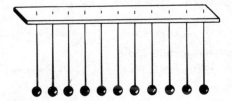

Using the equation

$$l\frac{d^2\psi}{dt^2} = -g\sin\psi$$

for the motion of a simple pendulum of length l which makes angle ψ with the vertical, and the equation

$$c^2\frac{\partial^2\psi}{\partial t^2} = \frac{\partial^2\psi}{\partial x^2}$$

for torsional waves along an elastic column, argue plausibly that the angle of your pins is governed by a sine–Gordon equation.

Seek to develop kinks and kink interactions with your pins and rubber strip (but do not let your expectations of a very good experimental model of the sine–Gordon equation bring you disappointment).

(Scott, 1970, §§2.7, 5.5)

Q8.3 *A resonant interaction.* Take the expression for f given in Q5.30, for the 2D KdV equation, and show that this generates a solution, u, even if $A_2 = 0$. Examine the nature of this new solution by considering: θ_1 fixed, $\theta_2 \to -\infty$; θ_2 fixed, $\theta_1 \to -\infty$; $\theta_3 = \theta_1 - \theta_2$ fixed, $\theta_1 \to +\infty$.

Introduce a parametrisation of the dispersion relation $k_i\omega_i = k_i^4 + 3l_i^2$ in the form

$$k_i = m_i + n_i, \qquad l_i = m_i^2 - n_i^2, \qquad \omega_i = 4(m_i^3 + n_i^3), \qquad i = 1,2,.$$

Hence show that $A_2 = 0$ if, for example, $m_1 = m_2$. Write $\theta_3 = k_3x + l_3y - \omega_3t + \alpha_3$ and show that, if $m_1 = m_2, n_3 = n_2$ and $m_3 = -n_1$, then $k_3\omega_3 = k_3^4 + 3l_3^2$.

[These definitions of ω_3, k_3 and l_3 (i.e. $\omega_3 = \omega_2 - \omega_1$, etc., and ω_i, k_i, l_i satisfying the dispersion relation for $i = 1,2,3$) are the conditions for a *resonant wave interaction*. Miles, 1977; Freeman, 1980.]

Q8.4 *The reduced Maxwell–Bloch equations.* In the theory of nonlinear optics, it has been found that an ultrashort pulse of light in a dielectric medium of two-level atoms is governed by the system of equations,

$$E_t = s, \qquad s_x = Eu + \mu r, \qquad r_x = -\mu s \qquad \text{and} \qquad u_x = -Es,$$

where E, s, u and r are functions of x and t, and μ is a constant.

Show that when $\mu = 0$ this system has solutions such that $u = \cos\phi$, where $E = \phi_x$, and deduce that ϕ satisfies a sine–Gordon equation.

Discuss the wave solutions of permanent form when $\mu \neq 0$. In particular, seeking solitary waves for which $E, r, s \to 0$ and $u \to -1$ as $x \to \pm\infty$, show that

$$E(x,t) = a\,\mathrm{sech}\,[\tfrac{1}{2}a\{x - 4t/(4\mu^2 + a^2) + \delta\}]$$

gives a solution for all a and δ.

[Lamb, 1971; also Gibbon *et al.*, (1973) used the method of inverse scattering to solve the initial-value problem for this system and to generate multisoliton solutions.]

*Q8.5 *Vibrations of a monatomic lattice.* A crystal may be modelled by a lattice of equal particles of mass m. The interaction of neighbouring particles is represented by nonlinear springs. It is given that it follows from Newton's second law of motion that

$$m\ddot{u}_n = 2f(u_n) - f(u_{n+1}) - f(u_{n-1})$$

for $n = 0, \pm 1, \pm 2, \ldots$, where u_n is the displacement of the nth particle from its position of equilibrium and f is the force function of the springs.

Linearise this system for small u_n, and show that there are travelling waves of the form

$$u_n = \varepsilon \exp(i\theta_n), \qquad \theta_n = \omega t - pn$$

provided that

$$m\omega^2 = -4\sin^2(\tfrac{1}{2}p)f'(0) > 0.$$

For the *Toda lattice* or *Toda chain*, take the force function $f(u) = -a(1 - e^{-bu})$ for some constants $a, b > 0$. To solve the system in this case, define s_n by $\dot{s}_n = f(u_n)$ and by the condition that $s_n = 0$ when $u_n = 0$, and seek a solution of the form $s_n = g(\theta_n)$. Deduce that

$$\frac{m\omega^2}{b} \frac{g''(\theta_n)}{a + \omega g'(\theta_n)} = g(\theta_n + p) + g(\theta_n - p) - 2g(\theta_n).$$

Show that there exist solitary-wave solutions of this differential-difference equation having the form $s_n = m\omega b^{-1}\tanh\theta_n$, if $m\omega^2 = ab\sinh^2 p$.

[Kittel, 1976, Chap. 4; Toda, 1967a, b. It can be shown (i) by use of elliptic functions that there are periodic nonlinear waves of permanent form, analogous to cnoidal waves; (ii) that the solitary waves above interact like solitons; and (iii) that the continuous limit of the discrete system is governed by a partial differential equation of the form in Q2.1(iv).]

*Q8.6 *Modons.* It is given that the motion of a thin layer of an incompressible inviscid fluid on a rapidly rotating sphere is governed approximately by a two-dimensional vorticity equation of the form

$$\zeta_t - \frac{1}{R^2}\psi_t + \beta\psi_x + \psi_x\zeta_y - \psi_y\zeta_x = 0,$$

where ψ is the stream function and $\zeta = \psi_{xx} + \psi_{yy}$ is the relative vorticity. A locally Cartesian frame fixed to the sphere is used, with the x-axis pointing eastwards and the y-axis northwards. The constant R is a certain length, called the radius of deformation, and β is the value of the northward derivative of the Coriolis parameter at the latitude where $y = 0$.

Seeking a wave of permanent form, show that if $\psi(x, y, t) = g(x - ct, y)$

for some function g and constant velocity c, then

$$g_{xx} + g_{yy} - g/R^2 + \beta y = G(g + cy)$$

for some differentiable function G of integration.

Show that if

$$g(r, \theta) = ac\sin\theta \times \left\{ \begin{array}{ll} \dfrac{q^2 J_1(kr/a)}{k^2 J_1(k)} - \left(1 + \dfrac{q^2}{k^2}\right)\dfrac{r}{a} & \text{for } r < a \\[3mm] -\dfrac{K_1(qr/a)}{K_1(q)} & \text{for } r > a \end{array} \right\},$$

where $x = r\cos\theta$ and $y = r\sin\theta$, a, k and q are positive constants, and J_1 is the Bessel function and K_1 the modified Bessel function, then

$$\zeta(r, \theta) = -a^{-1}cq^2\sin\theta \times \left\{ \begin{array}{ll} J_1(kr/a)/J_1(k) & \text{for } r < a \\ K_1(qr/a)/K_1(q) & \text{for } r > a \end{array} \right\}$$

Deduce that this gives a solution of the vorticity equation with

$$G(g + cy) = \left\{ \begin{array}{ll} \{\beta/c - (k^2 + q^2)/a^2\}(g + cy) & \text{for } r < a \\ \beta(g + cy)/c & \text{for } r > a \end{array} \right\}$$

if $q = a(R^{-2} + \beta/c)^{1/2}$.

Verify that for this solution ψ, ψ_r and ζ are continuous at $r = a$ if k is one of the (countable infinity of) roots of

$$\frac{K_2(q)}{qK_1(q)} = -\frac{J_2(k)}{kJ_1(k)}.$$

[These localised vortices are called *modons*. They are generalisations of well-known solutions for two-dimensional rotational flow (cf. Batchelor 1967, p. 535) which may be important in meteorology and oceanography. They seem not to be solitons because numerical calculations indicate that some do not retain their identities after interactions with one another. Stern, 1975; Larichev & Reznik, 1976; McWilliams & Zabusky, 1982.]

Q8.7 *The Born–Infeld equation.* Use the variational principle $\delta \int \mathscr{L} \, dx \, dt = 0$ with the Lagrangian density $\mathscr{L} = (1 - u_t^2 + u_x^2)^{1/2}$ to deduce that

$$(1 - u_t^2)u_{xx} + 2u_x u_t u_{xt} - (1 + u_x^2)u_{tt} = 0.$$

Show that $u(x, t) = f(x \pm t)$ gives two solutions for all twice differentiable functions f.

[Born & Infeld (1934) proposed a three-dimensional form of the equation to model a relativistic particle. See also Barbashov & Chernikov (1967).]

Answers and hints

The answer, if one is given, is designated by the prefix A: for example, the answer to Q1.1 is A1.1. In some cases a hint to the solution is included.

Chapter 1

A1.1 $u(x,t) = f(x - ct) + g(x + ct)$.

A1.2 $u(x,t) = \frac{1}{2}\{p(x - ct) + p(x + ct)\} + \dfrac{1}{2c}\displaystyle\int_{x-ct}^{x+ct} q(y)\,dy$.

A1.3 $q(x) = -c\,dp/dx$.

A1.4 Dispersion relation is $\omega = k - k^3 - ik^2$.

A1.5 First gives $\omega = k - k^3$; second gives $\omega = k/(1 + k^2)$ so that $\omega = k - k^3 + O(k^5)$ as $k \to 0$ and $\omega \sim 1/k$ as $k \to \infty$.

A1.6 $u(x,t) = \begin{cases} u_0(x - t)/(1 + u_0 t), & 0 \leqslant (x - t)/(1 + u_0 t) \leqslant 1 \\ u_0(2 - x + t)/(1 - u_0 t), & 1 \leqslant (x - t - 2u_0 t)/(1 - u_0 t) \leqslant 2 \\ 0, & \text{otherwise.} \end{cases}$

A1.7 See Fig. A1.

A1.8 $u(x,t) = \cos\{\pi(x - ut)\}$, $u_x = \pi \sin\{\pi(x - ut)\}/[1 - \pi t \sin\{\pi(x - ut)\}]$.

A1.9 $u(x,t) = f\{x - c(u)t\}$, $u_x = f'(\lambda)/\{1 + tc'(\lambda)f'(\lambda)\}$, where $\lambda = x - c(u)t$ is constant on each characteristic.

A1.10 $u(x,t) = \displaystyle\int_{-\infty}^{\infty} F(k)e^{i(kx - \omega t)}\,dk$, where $\omega = -k^3$ and $f(x) = \displaystyle\int_{-\infty}^{\infty} F(k)e^{ikx}\,dk$.

Remember that $\mathrm{Ai}(z) = \dfrac{1}{2\pi}\displaystyle\int_{-\infty}^{\infty} \exp\{i(kz + \tfrac{1}{3}k^3)\}\,dk$.

Fig. A.1 Sketch of A1.6. (a) $t = 0$, (b) $t = 1/u_0$.

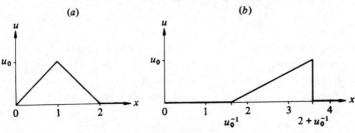

A1.11 $u(x, t) = -2\kappa^2 \operatorname{sech}^2 \{\kappa(x - 4\kappa^2 t) + A\}$, A arbitrary constant.

A1.13 $\lambda^2 = 1$.

A1.14 Use $c = (gh)^{1/2}\{1 - \frac{1}{6}k^2 h^2 + o(k^2 h^2)\}$ as $kh \to 0$ (for $c > 0$) and remember that

$$\frac{1}{2\pi} \int_{-\infty}^{\infty} e^{ikx} \, dk = \delta(x).$$

A1.15 $K(x) = c_0 \delta(x) + \dfrac{i\lambda c_0}{2\pi} \dfrac{d}{dx} \left(\displaystyle\int_0^{\infty} - \int_{-\infty}^0 \right) e^{ikx} \, dk$,

$$\int_{-\infty}^{\infty} K(x - \xi) u_\xi \, d\xi = c_0 u_x + i\lambda c_0 \int_{-\infty}^{\infty} (\operatorname{sgn} k) F(k) e^{ikx} \, dk, \quad \text{where}$$

$$F(k) = \frac{1}{2\pi} \int_{-\infty}^{\infty} u_{\xi\xi} e^{-ik\xi} \, d\xi.$$

Chapter 2

A2.1 (i) $f(\xi) = \frac{1}{2} u_0 \{1 - \tanh(\frac{1}{4} u_0 \xi + A)\}$;

(ii) $f(\xi) = \pm c^{1/2} \operatorname{sech}(c^{1/2}\xi + A)$ for all $c \geq 0$;

(iii) $f''(\xi) = \frac{1}{2} c \operatorname{sech}^2 (n c^{1/2}\xi + A)$;

(iv) $f(\xi) = f_0 + 6(c^2 - 1)^{1/2} \tanh \{\frac{1}{2}(c^2 - 1)^{1/2}\xi + A\}$,

where A and f_0 are arbitrary constants.

A2.2 Sketch the graphs of $(f')^2$ against f.

A2.3 Special solution obtained if $A = B = 0$.

A2.5 Unstable saddle point at $f' = f = 0$ if $c < -2\alpha$, and an unstable node if $c = -2\alpha$. At $f' = 0$, $f = 1$ there is a saddle point.

A2.6
$$f(\xi) = \begin{cases} A(1 - e^{-\lambda\xi}), & \xi \geq 0 \\ -\dfrac{A\lambda}{\mu}(1 - e^{\mu\xi}), & \xi \leq 0 \end{cases}$$

where $\lambda = (c - 1)^{1/2}, \mu = (c + 1)^{1/2}$ $(c > 1)$ and A is arbitrary. Solitary waves, for which the solution is sinusoidal in the middle and exponential in the tails, exist if $-1 < c < 1$; periodic waves exist if $c < -1$.

A2.7 (i) Compare the series in powers of m for both sides of the equation.

A2.8 $m = n = -\frac{1}{2}, -\frac{1}{2}\eta f + \frac{1}{2}f^2 = f', f(\eta) = -2\exp(-\frac{1}{4}\eta^2)/\int_{-\infty}^{\eta} \exp(-\frac{1}{4}y^2)\,dy$.

A2.10 $f''' + 6ff' - 2(f + 2\eta f') = 0$.

A2.11 Use $u(x, t) = t^{-1/2} f(xt^{-1/2})$; then $f'' - \frac{1}{2}i(\eta f)' + vf|f|^2 = 0$.

A2.12 Try $u(x, t) = c + f(x - \alpha t)$ and deduce that
$\frac{1}{2}(f')^2 + \frac{1}{2}(c + f)^4 - \frac{1}{2}\alpha f^2 = 2c^3 f + \frac{1}{2}c^4$. Hence $(g')^2 = (\alpha - 6c^2)g^2 - 4cg - 1$
if $g = 1/f$.

A2.13 Note that $\mathscr{H}(f_{xx}) = \{\mathscr{H}(f)\}_{xx}$. Solution requires $a = 1/c$ and $b = 4c$.

A2.15 First show that $v_t + 2v_x^3 + v_{xxx} = 0$. Express $\phi(x, t) = F(kx + mt + \alpha)/G(lx + nt + \beta)$
and take $F'' = -F, F^2 + (F')^2 = l^2, G'' = G, G^2 - (G')^2 = k^2$.

A2.16 Try $u(x, t) = a\exp(ia^2 t)(1 + A + iB)$ where $A = A(x, t), B = B(x, t)$ are both real.

A2.20 Write $\phi(x,t) = 4\arctan\{f(x)/g(t)\}$.

A2.22 As for A2.20, but for a different choice of f and g. If $F = \tan(\tfrac{1}{4}\phi)$ then

$$F(x,t) \sim \pm\lambda\exp\{\pm(x+\lambda t)/(1-\lambda^2)^{1/2}\} \mp \lambda\exp\{\mp(x-\lambda t)/(1-\lambda^2)^{1/2}\}$$

as $t \to \mp\infty$, and note that $\pm\lambda e^\theta = \pm\exp\{\theta + \tfrac{1}{2}\delta/(1-\lambda^2)^{1/2}\}$ where $\delta = -2(1-\lambda^2)^{1/2}\log\lambda$.

A2.24 One approach is to introduce X and then seek a solution
$$u(x,y,t) = f(X^2,y^2).$$

Chapter 3

A3.2 Form $\psi_n(L-\lambda)\hat{\psi} - \hat{\psi}(L+\kappa_n^2)\psi_n = 0$ and integrate from $x = -\infty$ to $+\infty$, where $L = -d^2/dx^2 + u$.

A3.3 Set $\mu = \lambda - U_0$. There is a continuous eigenfunction if $\mu \geqslant 0$ and none otherwise.

A3.4 (i) If $\lambda = k^2 > 0$ and $l^2 = U_0 - k^2 > 0$ then $\psi(x) = Ae^{lx} + Be^{-lx}$ for $0 < x < 1$ etc., and continuity of ψ, ψ' at $x = 0,1$ gives $a(1 \mp ik/l)e^{\pm l} = b(1 \pm ik/l)e^{ik} + (1 \mp ik/l)e^{-ik}$. If $\lambda = k^2 > 0$ and $k^2 - U_0 = m^2 > 0$, then as above with $l \to im$. If $k = \sqrt{U_0}$, then as above with $l \to 0$ and $a(k) = e^{-ik}/(1 - \tfrac{1}{2}ik)$.

(ii) If $\lambda = k^2 > 0$ then

$$a(k) = 2k^2\cos k/[\{2k^2 - ik(U_0 + U_1)\}e^{-ik} + iU_0U_1\sin k]$$

and

$$b(k) = \left\{\frac{iU_0}{2k} + \left(1 - \frac{iU_0}{2k}\right)e^{-2ik}\right\}a - e^{-2ik}.$$

If $\lambda = -\kappa_n^2 < 0$ then

$$(\kappa_n - \tfrac{1}{2}U_0)(\kappa_n - \tfrac{1}{2}U_1) = \tfrac{1}{4}U_0U_1\exp(-2\kappa_n).$$

Examine slope of $y = f(x) = (x - \tfrac{1}{2}U_0)(x - \tfrac{1}{2}U_1)$ and

$$y = g(x) = \tfrac{1}{4}U_0U_1e^{-2x} \qquad \text{at } x = 0.$$

A3.5 If $\lambda = k^2 > 0$ and $l^2 = k^2 - U_0 > 0$ then

$$\psi(x) = \begin{cases} ae^{-ikx}, & x < 0, \\ e^{-ilx} + be^{ilx}, & x > 0. \end{cases}$$

with $a = 2l/(l+k)$, $b = (l-k)/(l+k)$.

If $\lambda = k^2 > 0$ and $m^2 = U_0 - k^2 > 0$ then

$$\psi(x) = \begin{cases} A_-e^{-ikx} + A_+e^{ikx}, & x < 0 \\ Ce^{-mx} & x > 0 \end{cases}$$

with $A_\mp = \tfrac{1}{2}C(1 \mp im/k)$.

A3.6 Let $X = \beta x$, $\Lambda = \lambda/\beta^2$, $u_0 = U_0/\beta^2$.

A3.7 $\psi_1 = \tfrac{1}{4}\sqrt{3}S(5T^2 - 1)$, $\psi_2 = \tfrac{1}{2}\sqrt{15}S^2T$, $\psi_3 = \tfrac{1}{4}\sqrt{15}S^3$ where $S = \operatorname{sech} x$, $T = \tanh x$.

A3.8 (i) Set $\psi(x;k,\varepsilon) \sim \psi_0(x;k) + \varepsilon\psi_1(x;k)$.

(ii) Set $A(X;\varepsilon) \sim A_0(X) + \varepsilon A_1(X)$, $g(x, X;\varepsilon) \sim g_0(x, X) + \varepsilon g_1(x, X)$ and $\kappa = \varepsilon\hat{\kappa}(\varepsilon)$, $\hat{\kappa} \sim \kappa_0$.

A3.10 (i) Use $\Gamma(z)\Gamma(1 - z) = \pi/\sin \pi z$.

(ii) Use Stirling's formula, and $\sin \pi(1 - ik) = i \sinh \pi k$.

A3.11 If $L, L_z \to 0$ as $z \to -\infty$ then

$$L_{xx} - L_{zz} - u(x)L = 0, \qquad z < x,$$

and

$$u(x) = 2d\hat{L}/dx, \qquad \text{where } \hat{L}(x) = L(x, x).$$

A3.12 (i) If $k = i\kappa_n$ then $\psi_- \propto \psi_+$.

(ii) L.h.s. of equation (3.44) $\neq 0$ when $k = i\kappa_n$.

A3.13 (i) $K(x, z) = -e^{-(x+z)}/(1 + \frac{1}{2}e^{-2x})$;

(ii) $\phi(x, z) = -\frac{3}{4}xz$;

(iii) $K(x, z) = H(x + z)$ (see example (i), §3.4);

(iv) $\phi(x) = 1 + (2\cos x + \pi \sin x)/(1 - \frac{1}{4}\pi^2)$.

A3.14 $K_0 = 0$, $K_1 = e^{-(x+z)}$, $K_2 = -e^{-(x+z)}(1 - \frac{1}{2}e^{-2x})$, etc.

A3.15 N.B. Only the continuous spectrum is present,

$$u(x) = -2\beta\delta(x).$$

A3.16 $|A| = 1 + \sum_{i=1}^{3} E_i + \sum_{\langle i=1 \rangle}^{3} A_{ij}E_iE_j + \sum_{\langle i=1 \rangle}^{3} A_{ij}E_i$

where $E_i = c_i^2 \exp(-2\kappa_i x)/2\kappa_i$, $A_{ij} = (\kappa_i - \kappa_j)^2/(\kappa_i + \kappa_j)^2$ and $\langle\ \rangle$ denotes j chosen cyclically with respect to i.

Chapter 4

A4.1 Differentiate appropriately the Marchenko equation, and use integration by parts as required. For (iii), introduce the derivatives of (1) into (2).

A4.2 (i) Write $u(x, t) = (12t)^{-2/3}g(X, t)$ to give

$$12tg_t - 2g - 4Xg_X - 6gg_X + g_{XXX} = 0,$$

and then follow A4.1.

(ii) Let $F(x, z; t) = a(x, t)b(z, t)$ so that $a = \text{Ai}(x + A)$, $b = \text{Ai}(z + A)$, where A is an arbitrary constant.

A4.3 $F(x, z; t) = \exp\{-k(x + z) + 8k^3t + \alpha\} = g(x, z; t, k, \alpha)$. If $F(x, z; t) = g(x, z; t, k, \alpha) = g(x, z; t, l, \beta)$ then

$$u(x, t) = -2(k + l)^2\{(c_+ + \mu c_-)(c_+ + c_-/\mu) - (s_+ - s_-)^2\}/(c_+ + \mu c_-)^2$$

where

$$\binom{c}{s}_\pm = \binom{\cosh}{\sinh}\{(k \pm l)x - 4(k^3 \pm l^3)t\} \quad \text{and} \quad \mu = (k + l)/(k - l) \quad (k \neq l).$$

(Emerging solitons are of amplitudes $-2k^2$, $-2l^2$.)

A4.4 (i) $u(x, t) = -\frac{9}{2}\text{sech}^2\{\frac{3}{2}(x - 9t)\}$.

(ii) Three-soliton solution (see example (iii), §4.5 and A3.16).

(iii) In general, $b(k) \neq 0$. If $V \geqslant 0$ there are $1 + [2\sqrt{V}/\pi]$ solitons; if $V < 0$, none. If $\frac{1}{2}\{1 + 4V)^{1/2} - 1\}$ is integral then $b(k) = 0$ for all k.

A4.5 See A4.3 with $k > l$. Evaluate on $t = 0$, and examine u near $x = 0$. Two maxima (peaks) if $\sqrt{3} > k/l > 1$; only one maximum if $k/l \geqslant \sqrt{3}$. N.B. Profile is sech^2 if $k = 2$, $l = 1$.

A4.6 See A3.7 and A3.16. Let $\theta_n = 8n^3 - 2nx$. Then $u(x, t) \sim -2\,\text{sech}^2(x - 4t)$ as $t \to -\infty$ and $u(x, t) \sim -2\,\text{sech}^2(x + x_0 - 4t)$ as $t \to +\infty$ for fixed θ_1, where $x_0 = -\frac{1}{2}\log(A_{12}A_{31})$. Similarly for θ_2, θ_3 fixed.

A4.7 If $u = O(\alpha)$ then $K = O(\alpha)$ and so $F = O(\alpha)$, $\alpha \to 0$.

A4.8 Use the conserved density of Q5.2. The centre of mass has the x-coordinate $\int_{-\infty}^{\infty} xu\,dx / \int_{-\infty}^{\infty} u\,dx$.

A4.9 Form equation for L_n from that for K, and differentiate it to obtain $d^2 L_n/dx^2$. Examine behaviour as $x \to \infty$. Express L_n in terms of ψ_n in the equation for L_n, multiply by $(4n\psi_n - 2\psi'_n)$ and sum. Add this to $2\psi_n u_x$ expressed in terms of ψ_n and summed.

A4.10 Similar to A4.9, to give

$$u(x, t) = -\frac{2i}{\pi} \int_{-\infty}^{\infty} kb(k; t)\hat{\psi}^2(x; k, t)\,dk - 4\sum_{n=1}^{N} n\psi_n^2(x; t).$$

Chapter 5

A5.2 $(xu + 3tu^2)_t = (12tu^3 - 6tuu_{xx} + 3tu_x^2 + 3xu^2 - xu_{xx} + u_x)_x.$

A5.3 $u, u^2, u^4 + u_x^2$ are conserved densities.

A5.5 Remember that $|u|^2 = uu^*$.

A5.6 Note that $\mathcal{H}(u_x) = \{\mathcal{H}(u)\}_x$, etc.

A5.10 $u(x, t) \sim -\sum_{n=1}^{N} k_n^2 \text{sech}^2\{k_n(x - 4k_n^2 t - x_n)\}$ as $t \to \infty$, Let $f(x) = u(x, 0)$ then

$$\int_{-\infty}^{\infty} f(x)\,dx = \int_{-\infty}^{\infty} u(x, t)\,dx = -2\sum_{n=1}^{N} k_n$$

and

$$\int_{-\infty}^{\infty} \{f(x)\}^2\,dx = \int_{-\infty}^{\infty} \{u(x, t)\}^2\,dx = \frac{4}{3}\sum_{n=1}^{N} k_n^3, \text{etc.}$$

A5.11 Show that, if ψ_0 is an eigenvector of $L(0)$ with eigenvalue λ_0, then $U\psi_0$ is an eigenvector of $L(t)$ with eigenvalue λ_0. Show that $M = U_t\hat{U}$.

A5.13 Take $L = \begin{pmatrix} y & x \\ x & -y \end{pmatrix}$ and $M = \frac{1}{2}g\begin{pmatrix} 0 & -1 \\ 1 & 0 \end{pmatrix}$.

A5.15 $u_t + 210u^3u_1 - 105u_1^3 - 420uu_1u_2 - 42u_1u_2 - 105u^2u_3 - 21uu_3 - u_7 = 0$,
where $u_n = \partial^n u/\partial x^n$.

A5.18 Requires $A = -2v$.

A5.22 Show that, if $w = \log(f^*/f)$, then

$$\mathscr{H}(w_x) = \mathscr{H}\left\{\sum_{n=1}^{N}\left(\frac{1}{x-x_n^*} - \frac{1}{x-x_n}\right)\right\} = \sum_{n=1}^{N}\left(\frac{i}{x-x_n^*} + \frac{i}{x-x_n}\right)$$

$$= i\left(\frac{f_x}{f} + \frac{f_x^*}{f^*}\right).$$

A5.25 Show that

$$(f^2 - g^2)\{(D_x^2 - D_t^2 - 1)(f\cdot g)\} - 2fg\{(D_x^2 - D_t^2)(f\cdot f - g\cdot g)\} = 0.$$

A5.26 Show that

$$\{(iD_t + \beta D_x^2 + i\gamma D_x^3)(g\cdot f)\}/f^2 + 3i\{\alpha gg^* - \gamma D_x^2(f\cdot f)\}(fg_x - gf_x)/f^4$$
$$+ g\{\delta gg^* - \beta D_x^2(f\cdot f)\}/f^3 = 0.$$

A5.27 Write $D_x(D_t + D_x^3)(f\cdot f) = 0$ as $ff_{xt} - f_x f_t + f^2 F = 0$,
and differentiate to find F_x, F_{xx}. Form $f^2 F_{xx} + 3FD_x^2(f\cdot f)$ and $D_x^6(f\cdot f)$.
[Note that $f^2 F = ff_4 - 4f_1f_3 + 3f_2^2$, $f_n = \partial^n f/\partial x^n$.]

A5.28 (5.21) $u = -\partial^2(\log f)/\partial x^2$, $f = 1 + e^\theta$, $\theta = kx - \omega t + \alpha$, $\omega^2 = k^2 + k^4$;
    (5.22) $u = \frac{1}{2}i\partial\log(f^*/f)/\partial x = \lambda/\{\lambda^2(x - \lambda t + x_0)^2 + 1\}$,
            $f = i(x - \lambda t + x_0) + 1/\lambda$, λ real (cf. Q2.13);
    (5.23) $f = 1 + kt^{-1/3}\int_\xi^\infty A_i^2(y)\,dy$ where $\xi = x/(12t)^{1/3}$ and k is an arbitrary
    constant;
    (5.24) $f = 1 + e^\theta$, $\theta = kx + ly - \omega t + \alpha$, $\omega = k^3 + 3l^2/k$;
    (5.25) $f = 1$, $g = e^\theta$, $\theta = \frac{1}{2}\{a(x\pm t) + (x\mp t)/a\} + \alpha$;
    (5.26) $g = e^\theta$, $f = 1 + (\alpha/2\gamma)(k+k^*)^{-2}\exp(\theta + \theta^*)$,
            $\theta = kx + (i\beta k^2 - \gamma k^3)t + A$;
    (5.27) $f = 1 + e^\theta$, $\theta = kx - k^5t - k^3\tau + \alpha$.

A5.29 $f = |A|$ as given in A3.16.

A5.30 $A_2 = \dfrac{\{(k_1 - k_2)(\omega_2 - \omega_1) + (k_1 - k_2)^4 + 3(l_1 - l_2)^2\}}{\{(k_1 + k_2)(\omega_1 + \omega_2) - (k_1 + k_2)^4 - 3(l_1 + l_2)^2\}}$.

A5.31 $v(x, t) = f(x) + g(t)$ and then

$$u(x, t) = g(t) - f(x) + 2\log[-\sqrt{2}/\{F(x) + G(t)\}]$$

where $F(x) = \int^x e^{-f(x)}\,dx$, $G(t) = \int^t e^{g(t)}\,dt$.

A5.33 The four solutions are related by the theorem of permutability (see Fig. 5.1)
with $w_0 = u_1$, $w_1 = u_2$, $w_2 = u_3$, $w_{12} = u_4$, $\lambda_1 = a_1$, $\lambda_2 = a_2$. Form, and then
eliminate, the terms $(u_1 - u_4)_x$, $(u_1 - u_4)_t$.

A5.34 For example $u_4 = 4\arctan\left\{\dfrac{a_1 + a_2}{a_1 - a_2}\tan\left(\dfrac{u_2 - u_3}{4}\right)\right\}$ with any pair of
solutions u_2, u_3, say $\tan(u_n/4) = \exp(\theta_n)$, $\theta_n = a_nx + t/a_n + \alpha_n$ $(n = 2, 3)$, giving
the two-kink solution.

A5.35 Use equation (5.91), with $w_0 = 0$, to obtain

$$w_{123} = \left\{ \sum_{\langle i=1 \rangle}^{3} \lambda_i w_i(w_j - w_k) \right\} \Bigg/ \left\{ \sum_{\langle i=1 \rangle}^{3} \lambda_i(w_j - w_k) \right\}$$

where $\langle i = 1 \rangle$ denotes i, j, k taken cyclically. A three-soliton solution is obtained with $\lambda_1 > \lambda_2 > \lambda_3$, w_1 and w_3 regular solutions, and w_2 a singular solution. See Wahlquist & Estabrook (1973).

A5.37, 5.38 If the bilinear equation is $B(f \cdot f) = 0$, then form $f_1 f_1 B(f_2 \cdot f_2) - f_2 f_2 B(f_1 \cdot f_1) = 0$ and use appropriate exchange formulae.

A5.39 Form $\partial(1)/\partial x = 0$ and $\partial(1)/\partial t - \partial(2)/\partial x = 0$.

Chapter 6

A6.2 $W(\psi_-, \psi_+) = -b$, $W(\hat{\psi}_-, \psi_+) = \hat{a}$, $W(\hat{\psi}_-, \hat{\psi}_+) = \hat{b}$.

A6.3 Show that $W(\hat{\psi}_-, \psi_+)(\zeta) = W(\psi_-, \hat{\psi}_+)(-\zeta)$, $W(\hat{\psi}_-, \hat{\psi}_+)(\zeta) = \pm W(\psi_-, \psi_+)(-\zeta)$. If $r = \pm q^*$ then $\begin{pmatrix} \hat{\psi}_{1+} \\ \hat{\psi}_{2+} \end{pmatrix} = \begin{pmatrix} \pm \psi_{2+}^* \\ \psi_{1+}^* \end{pmatrix}$, $\begin{pmatrix} \hat{\psi}_{1-} \\ \hat{\psi}_{2-} \end{pmatrix} = \begin{pmatrix} \mp \psi_{2-}^* \\ -\psi_{1-}^* \end{pmatrix}$, and so $W(\hat{\psi}_-, \psi_+) = W(\psi_-^*, \hat{\psi}_+^*)$, $W(\hat{\psi}_-, \hat{\psi}_+) = \pm W(\psi_-^*, \psi_+^*)$, i.e. $\hat{a} = a^*$, $\hat{b} = \mp b^*$.

A6.4 If $r = \pm q$ then $\hat{f} = \mp f$; if $r = \pm q^*$ then $\hat{f} = \mp f^*$, $K_{21} = \pm K_{12}^*$, $K_{11}^* = K_{22}$ and $K_{21} + f + \int_x^{\infty} K_{22} f \, dy = 0$; $K_{11} + \int_x^{\infty} K_{12} f \, dy = 0$.

A6.6 Conserved densities are
(i) q^2, $q_x^2 + q^4$, $q_{xx}^2 + 11 q^2 q_x^2 + 2q^6$;
(ii) $|q|^2$, $q^* q_x - q q_x^*$, $|q_x|^2 - \frac{1}{2}|q|^4$.

A6.7 $c = -4i$.

A6.8 See Q2.4: $q \to m \, e^{i(\theta + nt)}$ as $|x| \to \infty$.

A6.10 Operate only in $z < x$.

A6.15 $L = L' = \begin{pmatrix} I & 0 \\ 0 & 0 \end{pmatrix}$, $M = \begin{pmatrix} 0 & -q & 0 \\ -r & 0 & \\ & B & 0 \end{pmatrix}$, $N = \begin{pmatrix} -1 & 0 & 0 \\ 0 & 1 & \\ & 0 & I \end{pmatrix}$, $M' = \begin{pmatrix} 0 & A^{-1}B^{-1} \\ B & 0 \end{pmatrix}$, $N' = \begin{pmatrix} 0 & 0 \\ 0 & I \end{pmatrix}$.

A6.18 $u(x, t) = 4 \arctanh(e^{\lambda x + t/\lambda + \alpha})$.

A6.19 $u(x, t) = \dfrac{\{E_1 + E_2 + A^2(E_1^2 E_2/4k_1^2 + E_1 E_2^2/4k_2^2)\}}{(1 + E_1^2/4k_1^2 + E_2^2/4k_2^2 + BE_1 E_2 + A^4 E_1^2 E_2^2/16k_1^2 k_2^2)}$

where $E_i = \exp(\theta_i)$, $\theta_i = k_i x - k_i^3 t + \alpha_i$, $A = (k_1 - k_2)/(k_1 + k_2)$ and $B = 2/(k_1 + k_2)^2$. The solitary-wave solution is $u = k_1 \,\text{sech}\, \theta_1$.

A6.20 $u = -4\mu \,\text{sech}\, \theta \left\{ \dfrac{\cos \phi - (\mu/\lambda) \sin \phi \tanh \theta}{1 + (\mu/\lambda)^2 \sin^2 \phi \,\text{sech}^2\, \theta} \right\}$ where $\zeta_1 = \lambda + i\mu$, $\theta = 2\lambda x + 8\lambda(\lambda^2 - 3\mu^2)t + \alpha$, $\phi = 2\mu x + 8\mu(3\lambda^2 - \mu^2)t + \beta$.

A6.21 See Q2.21.

A6.22 Take, e.g., $\lambda_1 > -\lambda_2 > 0$. The solution for u_x represents the collision of two hump-shaped solitons: cf. KdV equation. (You may take $\delta_1 = \delta_2 = 0$ without loss of generality.)

A6.23 (i) E.g. $\lambda_1 > -\lambda_2 > 0$. (ii) E.g. $\lambda_1 > \lambda_2 > 0$. See §8.1 for some interactions in a laboratory frame.

Chapter 7

A7.1 (i) $w = A(z-1)^{1/2}$;

(ii) $w = A/(z - z_0) + B$;

(iii) $w = A \log(z - z_0) + B$.

A7.2 For the P-I equation, $w(z) \sim (z - z_0)^{-2}$ as $z \to z_0$.

A7.3 (i) $d^2w/d\eta^2 = 6w^2 + \eta$, $\eta = z - \frac{1}{24}$;

(ii) $d^2w/d\eta^2 = 2w^3 + \eta w$, $\eta = z^{1/3}$.

A7.5 $w'' = (w')^2/w - w'/z + \frac{1}{2}(w^2 - 1)/z$.

A7.6 $-\lambda F + F'' + \nu F|F|^2 = 0$, then $F \sim A(x - x_0)^{-1}$ as $x \to x_0$.

A7.7 If $N = 3$, then $2F + 4zF' + (z^2 - 1)F'' = F^3$ where $z = x/t$ (and $m = -1$), and so $F \sim A(z - z_0)^{-3}$ as $z \to z_0$.

A7.8 Let $u = F(z)$, $z = xt$, then $zF'' + F' = \frac{1}{2}(e^F - e^{-F})$. Transform $w = e^F$.

A7.9 $n = -\frac{2}{3}$, $p = -\frac{1}{3}$, $q = -\frac{4}{3}$, $\lambda = \frac{1}{12}$, then

$$F^{\mathrm{iv}} - 6FF'' - \tfrac{1}{3}zF'' - \tfrac{1}{2}F' - 6(F')^2 = 0$$

where $z = xt^{-1/3} + (1/12)y^2t^{-4/3}$. If $F = v^2$ then $v'' = v^3 + (1/12)zv$.

Chapter 8

A8.1 In linear theory, $A(x, t) = e^{i\{(k - K)x - (\omega - \Omega)t\}}$

$$\approx \exp\left[i\left\{(k - K)x - (k - K)f'(K)t + \tfrac{1}{2}(k - K)^2 f''(K)t\right\}\right].$$

A8.3 $A_2 = (m_1 - m_2)(n_1 - n_2)/\{(m_1 + n_2)(m_2 + n_1)\}$.

A8.4 $-\frac{1}{2}(E')^2 = \frac{1}{8}E^4 + \frac{1}{2}(\mu^2 + A/c)E^2 + \mu BE/c + D/c$ where $E = E(x - ct)$; for solitary waves $B = D = 0$, $A = -1$.

A8.5 After linearisation $m\ddot{u}_n = f'(0)(2u_n - u_{n+1} - u_{n-1})$. For nonlinear equation, $\ddot{s}_n = -bm^{-1}(a + \dot{s}_n)(2s_n - s_{n+1} - s_{n-1})$ and try $s_n = A \tanh \theta_n$.

A8.6 Vorticity equation becomes
$\partial(g + cy, \zeta - g/R^2 + \beta y)/\partial(x, y) = 0$; therefore $\zeta - g/R^2 + \beta y = G(g + cy)$. Here
$\zeta = (rg_r)_r/r + g_{\theta\theta}/r^2 = (rg_r)_r/r - g/r^2$.

A8.7 Use Euler–Lagrange equation.

Bibliography and author index

The numbers in square brackets following each entry give the pages of this book on which the entry is cited.

Ablowitz, M. J., Kaup, D. J., Newell, A. C. & Segur, H. (1974). The inverse scattering transform – Fourier analysis for nonlinear problems. *Stud. Appl. Math.*, **53**, 249–315. [127]

Ablowitz, M. J., Kruskal, M. D. & Ladik, J. F. (1979). Solitary wave collisions. *SIAM J. Appl. Math.*, **36**, 428–37. [187]

Ablowitz, M. J., Ramani, A. & Segur, H. (1978). Nonlinear evolution equations and ordinary differential equations of Painlevé type. *Lett. Nuovo Cim.*, **23**, 333–8. [172]

—— (1980a, b). A connection between nonlinear evolution equations and ordinary differential equations of P-type I, II. *J. Math. Phys.*, **21**, 715–21, 1006–15. [172]

Ablowitz, M. J. & Segur, H. (1977). Exact linearization of a Painlevé transcendent. *Phys. Rev. Lett.*, **38**, 1103–06. [174]

—— (1981). *Solitons and the Inverse Scattering Transform*. SIAM Studies in Applied Mathematics, Vol. 4, Philadelphia. [33, 37, 61, 86, 118, 143, 162, 187, 200]

Abramowitz, M. & Stegun, I. A. (ed.) (1964). *Handbook of Mathematical Functions*. Washington: Nat. Bureau of Standards. (Also New York: Dover, 1965.) [33, 60, 83]

Airault, H. (1979). Rational solutions of Painlevé equations. *Stud. Appl. Math.*, **61**, 33–54. [173]

Anker, D. & Freeman, N. C. (1978a). On the soliton solutions of the Davey–Stewartson equation for long waves. *Proc. Roy. Soc. A*, **360**, 529–40. [167]

—— (1978b). Interpretation of three-soliton interactions in terms of resonant triads. *J. Fluid Mech.*, **87**, 17–31. [196]

Aris, R. (1975). *The Mathematical Theory of Diffusion and Reaction in Permeable Catalysts*. Oxford University Press (2 vols.). [34]

Barbashov, B. M. & Chernikov, N. A. (1967). Solution of the two plane wave scattering problems in a nonlinear scalar field theory. *Sov. Phys. JETP*, **24**, 437–42. [204]

Batchelor, G. K. (1967). *An Introduction to Fluid Dynamics*. Cambridge University Press. [204]

Bell, E. T. (1937). *Men of Mathematics*. New York: Simon & Schuster. (Also Penguin Books, 1965, 2 vols.) [33]

Benjamin, T. B. (1967). Internal waves of permanent form in fluids of great depth. *J. Fluid Mech.*, **29**, 559–92. [19, 36]

—— (1972). The stability of solitary waves. *Proc. Roy. Soc. A*, **328**, 153–83. [26]

214   *Bibliography and author index*

Benjamin, T. B., Bona, J. L. & Mahony, J. J. (1972). Model equations for long waves in nonlinear dispersive systems. *Phil. Trans. Roy. Soc. A*, **227**, 47–78. [119]

Benney, D. J. (1966). Long non-linear waves in fluid flows. *J. Math. & Phys.*, **45**, 52–63. [17]

Ben-yu, G. & Manoranjan, V. S. (1985). A spectral method for solving the RLW equation. *IMA J. Numer. Analysis*, **5**, 307–18. [186]

Berezin, Yu. A. & Karpman, V. I. (1967). Nonlinear evolution of disturbances in plasmas and other dispersive media. *Sov. Phys. JETP*, **24** (5), 1049–56. [121]

Bhatnagar, P. L. (1979). *Nonlinear Waves in One-dimensional Dispersive Systems*. Oxford: Clarendon Press. [199, 200]

Bluman, G. W. & Cole, J. D. (1974). *Similarity Methods for Differential Equations*. New York: Springer-Verlag. [16]

Bona, J. L., Pritchard, W. G. & Scott, L. R. (1985). Numerical schemes for a model for nonlinear dispersive waves. *J. Computat. Phys.*, **60**, 167–86. [186]

Born, M. & Infeld, L. (1934). Foundations of the new field theory. *Proc. Roy. Soc. A*, **144**, 425–51. [204]

Boussinesq, J. (1871). Théorie de l'intumescence liquid appelée onde solitaire ou de translation, se propageant dans un canal rectangulaire, *Comptes Rendus Acad. Sci. (Paris)*, **72**, 755–9. [8]

Burgers, J. M. (1948). A mathematical model illustrating the theory of turbulence. *Adv. Appl. Mech.*, **1**, 171–99. [122]

Byrd, P. F. & Friedman, M. D. (1971). *Handbook of Elliptic Integrals for Engineers and Physicists*. 2nd edn. New York: Springer-Verlag. [33]

Calogero, F. & Degasperis, A. (1982). *Spectral Transform and Solitons I*. Amsterdam: North-Holland. [61, 118]

Candler, S. & Johnson, R. S. (1981). On the asymptotic solution of the perturbed KdV equation using the inverse scattering transform. *Phys. Lett.*, **86A**, 337–40. [180]

Cole, J. D. (1951). On a quasi-linear parabolic equation occurring in aerodynamics. *Quart. Appl. Math.*, **9**, 225–36. [122]

Coleman, S. (1975). Quantum sine–Gordon equation as the massive Thirring model. *Phys. Rev. D*, **11**, 2088–97. [199]

Courant, R. & Hilbert, D. (1953). *Methods of Mathematical Physics* (2 Vols.). New York: Wiley-Interscience. [60]

Craik, A. D. D. (1985). *Wave Interactions and Fluid Flows*. Cambridge University Press. [200]

Davey, A. & Stewartson, K. (1974). On three-dimensional packets of surface waves. *Proc. Roy. Soc. A*, **338**, 101–10. [167]

Davis, R. E. & Acrivos, A. (1967). Solitary internal waves in deep water. *J. Fluid Mech.*, **29**, 593–607. [19, 36]

Dodd, R. K., Eilbeck, J. C., Gibbon, J. D. & Morris, H. C. (1982). *Solitons and Nonlinear Wave Equations*. London: Academic Press. [61, 86, 162, 187]

Drazin, P. G. (1963). On one-dimensional propagation of long waves. *Proc. Roy. Soc. A*, **273**, 400–11. [62]

—— (1977). On the stability of cnoidal waves. *Quart. J. Mech. Appl. Math.*, **30**, 91–105. [26]

Dryuma, V. S. (1974). Analytic solution of the two-dimensional Korteweg-de Vries (KdV) equation. *Sov. Phys. JETP Lett.*, **19**, 387–8. [122]

Eilbeck, J. C. & McGuire, G. R. (1975). Numerical study of the regularized long wave equation, I: Numerical methods. *J. Computat. Phys.*, **19**, 43–57. [186]

Eisenhart, L. P. (1909). *A Treatise on the Differential Geometry of Curves and Surfaces*. Boston, Mass.: Ginn. (Also New York: Dover (1960).) [115]

Enneper, A. (1870). Über asymptotische Linien. *Nachr. Königl. Ges. Wiss. Göttingen*, 493–511. [109]

Faddeev, L. D. (1958). On the relation between the S-matrix and potential for the one-dimensional Schrödinger operator. *Dokl. Akad. Nauk SSSR*, **121**, 63–6. (Also *Amer. Math. Soc. Transl.* (2) (1967) **65**, 139–66.) [40]

Fermi, E., Pasta, J. & Ulam, S. M. (1955). Studies in nonlinear problems. *Tech. Rep.*, **LA-1940**, Los Alamos Sci. Lab. (Also in *Nonlinear Wave Motion*, (1974) ed. A. C. Newell, Providence, R. I. Amer. Math. Soc.; and in *Collected Papers of Enrico Fermi*, Vol. II, 1965, pp. 978–88, Chicago University Press.) [14]

Fife, P. C. (1979). Mathematical Aspects of Reacting and Diffusing Systems. *Springer Lecture Notes in Biomathematics*, Vol. 28 (Heidelberg: Springer-Verlag). [34]

Fisher, R. A. (1937). The wave of advance of advantageous genes. *Ann. Eugenics*, **7**, 355–69. (Also *Collected Papers*, Vol. IV, 69–83, University of Adelaide.) [34]

Fokas, A. S. & Ablowitz, M. J. (1981). Linearization of the Korteweg–de Vries and Painlevé II equations. *Phys. Rev. Lett.*, **47** (16), 1096–1100. [174]

Fornberg, B. & Whitham, G. B. (1978). A numerical and theoretical study of certain nonlinear wave phenomena. *Phil. Trans. Roy. Soc. A*, **289**, 373–404. [185, 187]

Forsyth, A. R. (1906). *Theory of Differential Equations, Part IV – Partial Differential Equations*, Vol. VI. Cambridge University Press. (Also New York: Dover, 1959.) [6, 122]

Freeman, N. C. (1980). Soliton interactions in two dimensions. *Adv. Appl. Mech.*, **20**, 1–37. [38, 122, 196, 202]

Freeman, N. C. & Johnson, R. S. (1970). Shallow water waves on shear flows. *J. Fluid Mech.*, **42**, 401–9. [16]

Frenkel, J. & Kontorova, T. (1939). On the theory of plastic deformation and twinning. *Fiz. Zhurnal*, **1**, 137–49. [199]

Garabedian, P. R. (1964). *Partial Differential Equations*. New York: Wiley. [50]

Gardner, C. S., Greene, J. M., Kruskal, M. D. & Miura, R. M. (1967). Method for solving the Korteweg–de Vries equation. *Phys. Rev. Lett.*, **19**, 1095–7. [86]

—— (1974). Korteweg–de Vries equation and generalizations. VI. Methods for exact solution. *Comm. Pure Appl. Math.*, **27**, 97–133. [86]

Gel'fand, I. M. & Fomin, S. V. (1963). *Calculus of Variations*. Englewood Cliffs, N J: Prentice Hall. [118]

Gel'fand, I. M. & Levitan, B. M. (1951). On the determination of a differential equation from its spectral function. *Izv. Akad. Nauk SSSR Ser. Mat.*, **15**, 309–66. (Also *Amer. Math. Soc. Transl.* (2), **1**, 253–304.) [60]

Gibbon, J. D., Caudrey, P. J., Bullough, R. K. & Eilbeck, J. C. (1973). An *N*-soliton solution of a nonlinear optics equation derived by a general inverse method. *Lett. Nuovo Cim.*, **8**, 775–9. [202]

Gibbon, J. D., James, I. N. & Moroz, I. M. (1979). An example of soliton behaviour in a rotating baroclinic fluid. *Proc. Roy. Soc. A*, **367**, 219–37. [199]

Gordon, W. (1926). Der Comptoneffekt der Schrödingerschen Theorie. *Zeit. für Phys.*, **40**, 117–33. [198]

Greig, I. S. & Morris, J. L. (1976). A hopscotch method for the Korteweg–de Vries equation. *J. Computat. Phys.*, **20**, 64–80. [184]

Hammack, J. L. & Segur, H. (1974). The Korteweg–de Vries equation and water waves Part 2: Comparison with experiments. *J. Fluid Mech.*, **65**, 289–314. [17]

Hirota, R. (1971). Exact solution of the Korteweg–de Vries equation for multiple collisions of solitons. *Phys. Rev. Lett.*, **27**, 1192–4. [102]

(1973a). Exact envelope-soliton solutions of a nonlinear wave equation. *J. Math. Phys.*, **14**, 805–9. [124]

—— (1973b). Exact N-soliton solution of the wave equation of long waves in shallow water and in nonlinear lattices. *J. Math. Phys.*, **14**, 810–4. [105, 168]

Hopf, E. (1950). The partial differential equation $u_t + uu_x = \mu u_{xx}$. *Comm. Pure Appl. Math.*, **3**, 201–30. [122]

Ince, E. L. (1927). *Ordinary Differential Equations*. London: Longmans, Green. (Also New York: Dover, 1956) [18, 32, 33, 60, 170, 171]

Johnson, R. S. (1973). On the development of a solitary wave moving over an uneven bottom. *Proc. Camb. Phil. Soc.*, **73**, 183–203. [16, 175]

—— (1979). On the inverse scattering transform, the cylindrical K–dV equation and similarity solutions. *Phys. Lett.*, **72A**, (3), 197–9. [174]

—— (1980). Water waves and Korteweg–de Vries equations. *J. Fluid Mech.*, **97**, 701–19. [16, 35, 122]

Kadomtsev, B. B. & Petviashvili, V. I. (1970).On the stability of solitary waves in weakly dispersing media. *Sov. Phys. Dokl.*, **15**, 539–41.[38, 122, 199]

Kakutani, T. & Ono, H. (1969). Weak non-linear hydromagnetic waves in a cold collision-free plasma. *J. Phys. Soc. Japan*, **26**, 1305–18. [198]

Karpman, V. I. & Maslov, E. M. (1978). Structure of tails produced under the action of perturbations on solitons. *Sov. Phys. JETP*, **48**, 252–9. [180]

Kaup, D. J. & Newell, A. C. (1978). Solitons as particles, oscillators and in slowly changing media: a singular perturbation theory. *Proc. Roy. Soc. A*, **361**, 413–46. [180]

Kay, I. & Moses, H. E. (1956). The determination of the scattering potential from the spectral measure function, III. Calculation of the scattering potential from the scattering operator for the one-dimensional Schrödinger equation. *Nuovo Cim.* (10), **3**, 276–304. [60]

Kevorkian, J. & Cole, J. D. (1981). *Perturbation Methods in Applied Mathematics*. New York: Springer-Verlag. [16]

Kittel, C. (1976). *Introduction to Solid State Physics*. 5th edn. New York: Wiley. [203]

Klein, O. (1927). Elektrodynamik und Wellenmechanik vom Standpunkt des Korrespondenzprinzips. *Zeit. für Phys.*, **41**, 407–42. [198]

Kline, M. (1972). *Mathematical Thought from Ancient to Modern Times*. New York: Oxford University Press. [33]

Knickerbocker, C. J. & Newell, A. C. (1980). Shelves and the Korteweg–de Vries equation. *J. Fluid Mech.*, **98**, 803–18. [180]

Kolmogorov, A. N., Petrovsky, I. G. & Piscounov, N. S. (1937). Study of the diffusion equation for growth of a quantity of a substance and application to a biological problem. *Bull. Moscow State Univ.*, **A1**, (6). [34]

Konno, K. & Wadati, M. (1975). Simple derivation of Bäcklund transformation from the Riccati form of the inverse method. *Progr. Theoret. Phys.*, **53**, 1652–6. [126]

Korteweg, D. J. & de Vries, G. (1895). On the change of form of long waves advancing in a rectangular canal, and on a new type of long stationary waves. *Phil. Mag.* (5), **39**, 422–43. [9, 12, 25]

Lamb, G. L., Jr. (1971). Analytical descriptions of ultrashort optical pulse propagation in a resonant medium. *Rev. Modern Phys.*, **43**, 99–124. [119, 202]

—— (1980). *Elements of Soliton Theory*. New York: Wiley-Interscience. [61, 118, 126, 162]

Landau, L. D. & Lifshitz, E. M. (1977). *Quantum Mechanics, Non-relativistic Theory*. 3rd. edn. London: Pergamon. [60, 62]

Larichev, V. D. & Reznik, G. M. (1976). Two-dimensional Rossby soliton: an exact solution. *Rep. USSR Acad. Sci.*, **231** (5) (in Russian). (Also in *POLYMODE News*, **19**, 3, 6 (in English).) [204]

Lax, P. D. (1968). Integrals of nonlinear equations of evolution and solitary waves. *Comm. Pure Appl. Math.*, **21**, 467–490. [87, 97, 118, 121, 190]

Lee, T. D. (1981). *Particle Physics and Introduction to Field Theory*. Chur, Switzerland: Harwood Academic. [200]

Lighthill, M. J. (1956). Viscosity effects in sound waves of finite amplitude. In *Surveys in Mechanics*, ed. G. K. Batchelor & R. M. Davies, pp. 250–351. Cambridge University Press. [122]

—— (1978). *Waves in Fluids*. Cambridge University Press. [16]

Ma, Y.-C. (1979). The perturbed plane-wave solution of the cubic Schrödinger equation. *Stud. Appl. Math.*, **60**, 43–58. [36]

McLeod, J. B. & Olver, P. J. (1983). The connection between partial differential equations soluble by inverse scattering and ordinary differential equations of Painlevé type. *SIAM J. Math. Anal.*, **14**, 488–506. [172]

McWilliams, J. C. & Zabusky, N. J. (1982). Interactions of isolated vortices I: Modons colliding with modons. *Geophys. Astrophys. Fluid Dyn.*, **19**, 207–27. [204]

Marchenko, V. A. (1955). On the reconstruction of the potential energy from phases of the scattered waves. *Dokl. Akad. Nauk SSSR*, **104**, 695–8. [60]

Matsuno, Y. (1984). *Bilinear Transformation Method*. Orlando, Fla: Academic Press. [103, 118]

Maxon, S. & Viecelli, J. (1974). Cylindrical solitons. *Phys. Fluids*, **17**, 1614–6. [35]

Miles, J. W. (1977). Resonantly interacting solitary waves. *J. Fluid Mech.*, **79**, 171–9. [202]

—— (1978a). An axisymmetric Boussinesq wave. *J. Fluid Mech.*, **84**, 181–91. [35]

—— (1978b). On the evolution of a solitary wave for very weak nonlinearity. *J. Fluid Mech.*, **87**, 773–83. [62]

Miura, R. M. (1976). The Korteweg–de Vries equation: a survey of results. *SIAM Rev.*, **18**, 412–59. [86]

Miura, R. M., Gardner, C. S. & Kruskal, M. D. (1968). Korteweg–de Vries equation and generalizations. II. Existence of conservation laws and constants of motion. *J. Math. Phys.*, **9**, 1204–9. [92, 120]

Morse, P. M. & Feshbach, H. (1953). *Methods of Theoretical Physics, Part II*. New York: McGraw-Hill. [62]

Newell, A. C. (1985). *Solitons in Mathematics and Physics*. Philadelphia, Pa: SIAM. [86, 162, 187]

Novikov, S., Manakov, S. V., Pitaevskii, L. P. & Zakharov, V. E. (1984). *Theory of Solitons: The Inverse Scattering Method*. New York: Consultants Bureau. [162]

Olver, P. J. (1979). Euler operators and conservation laws of the BBM equation. *Math. Proc. Camb. Phil. Soc.*, **85**, 143–60. [119]

Ono, H. (1975). Algebraic solitary waves in stratified fluids. *J. Phys. Soc. Japan*, **39**, 1082–91. [19, 36]

Peregrine, D. H. (1966). Calculations of the development of an undular bore. *J. Fluid Mech.*, **25**, 321–30. [119, 186]

—— (1983). Water waves, nonlinear Schrödinger equations and their solutions. *J. Austral. Math. Soc. Ser. B*, **25**, 16–43. [36]

Perring, J. K. & Skyrme, T. H. R. (1962). A model unified field equation. *Nucl. Phys.*, **31**, 550–5. [186, 187, 192, 199]

Rajaraman, R. (1982). *Solitons and Instantons*. Amsterdam: North-Holland. [200]

Rayleigh, Lord (1876). On waves. *Phil. Mag.* (5), **1**, 257–79. (Also *Sci. Papers* Vol. **I**, 251–71.) [8]

Rebbi, C. (1979). Solitons. *Sci. American*, **240** (2), 76–91. [200]

Redekopp, L. G. & Weidman, P. D. (1978). Solitary Rossby waves in zonal shear flows and their interactions. *J. Atmos. Sci.*, **35**, 790–804. [17]

Rogers, C. & Shadwick, W. F. (1982). *Bäcklund Transformations and Their Applications.* Vol. 161 in the series *Mathematics in Science and Engineering.* New York: Academic Press. [118]

Russell, J. S. (1844). Report on waves. *Rep. 14th Meet. Brit. Assoc. Adv. Sci., York*, 311–90. London: John Murray. [7, 12, 13]

Sanz-Serna, J. M. (1982). An explicit finite-difference scheme with exact conservation properties. *J. Computat. Phys.*, **47**, 199–210. [184]

Sawada, K. & Kotera, T. (1974). A method for finding N-soliton solutions of the K. d. V. equation and K. d. V.-like equation. *Progr. Theoret. Phys.*, **51**, 1355–67. [105]

Scott, A. C. (1970). *Active and Nonlinear Wave Propagation in Electronics.* New York: Wiley-Interscience. [202]

Segur, H. (1973). The Korteweg–de Vries equation and water waves: solutions of the equation. Part I. *J. Fluid Mech.*, **59**, 721–36. [83]

Shabat, A. B. (1973). On the Korteweg–de Vries equation. *Sov. Math. Dokl.*, **14**, 1266–9. [127]

Stern, M. E. (1975). Minimal properties of planetary eddies. *J. Marine Res.*, **33**, 1–13. [204]

Stewartson, K. & Stuart, J. T. (1971). A non-linear instability theory for a wave system in plane Poiseuille flow. *J. Fluid Mech.*, **48**, 529–45. [198]

Stoker, J. J. (1957). *Water Waves.* New York: Interscience. [32]

Taha, T. R. & Ablowitz, M. J. (1984a). Analytical and numerical aspects of certain nonlinear evolution equations. I. Analytical. *J. Computat. Phys.*, **55**, 192–202. [187]

—— (1984b). Analytical and numerical aspects of certain nonlinear evolution equations. II. Numerical, nonlinear Schrödinger equation. *J. Computat. Phys.*, **55**, 203–30. [186]

Toda, M. (1967a). Vibration of a chain with nonlinear interaction. *J. Phys. Soc. Japan*, **22**, 431–6. [203]

—— (1967b). Wave propagation in anharmonic lattices. *J. Phys. Soc. Japan*, **23**, 501–6. [203]

Vainshtein, A. I., Zakharov, V. I., Novikov, V. A. & Shifman, M. A. (1982). ABC of instantons. *Sov. Phys. Usp.*, **25**, 195–215. [200]

van Wijngaarden, L. (1968). On the equations of motion for mixtures of liquid and gas bubbles. *J. Fluid Mech.*, **33**, 465–74. [17]

Wadati, M. (1974). Bäcklund transformation for solutions of the modified Korteweg–de Vries equation. *J. Phys. Soc. Japan*, **36**, 1498. [126]

Wahlquist, H. D. & Estabrook, F. B. (1973). Bäcklund transformation for solutions of the Korteweg–de Vries equation. *Phys. Rev. Lett.*, **31**, 1386–90. [112, 118, 211]

Washimi, H. & Taniuti, T. (1966). Propagation of ion-acoustic solitary waves of small amplitude. *Phys. Rev. Lett.*, **17**, 996–8. [17]

Weiss, J., Tabor, M. & Carnevale, G. (1983). The Painlevé property for partial differential equations. *J. Math. Phys.*, **24**, 522–6. [172]

Whitham, G. B. (1974). *Linear and Nonlinear Waves.* New York: Wiley. [16, 32]

Wiegel, R. L. (1960). A presentation of cnoidal wave theory for practical application. *J. Fluid Mech.*, **7**, 273–86. [33]

Zabusky, N. J. (1967). A synergetic approach to problems of nonlinear dispersive wave propagation and interaction. In *Nonlinear Partial Differential Equations*, ed. W. F. Ames, pp. 223–58. New York: Academic Press. [33, 35, 198]

Zabusky, N. J. & Kruskal, M. D. (1965). Interactions of 'solitons' in a collisionless plasma and the recurrence of initial states. *Phys. Rev. Lett.*, **15**, 240–3. [14, 180, 184]

Zakharov, V. E. & Shabat, A. B. (1972). Exact theory of two-dimensional self-focusing and one-dimensional self-modulation of waves in nonlinear media. *Sov. Phys. JETP*, **34**, 62–9. [127, 164]

—— (1973). Interaction between solitons in a stable medium. *Sov. Phys. JETP*, **37**, 823–8. [34, 35]

—— (1974). A scheme for integrating the nonlinear equations of mathematical physics by the method of the inverse scattering problem I. *Funct. Anal. Appl.*, **8**, 226–35. [127, 144, 162]

Motion picture index

Various films about solitons have been made. Some 16 mm ones, which may be loaned or bought, are listed below. It is very instructive to see, in particular, computer-made animations of solitons and their interactions.

Eilbeck, J. C. (F1973). *Solitons and Bions in Non-linear Optics.* Sound, colour, 6 min.

Eilbeck, J. C. (F1975). *Numerical Solutions of the Regularized Long-Wave Equation.* Silent, colour, 3 min.

Eilbeck, J. C. (F1977a). *Boomerons.* Silent, colour, 6 min.

Eilbeck, J. C. (F1977b). *Zoomerons.* Silent, colour, 6 min.

Eilbeck, J. C. (F1978). *Two-dimensional Solitons.* Silent, colour, 4 min.

Eilbeck, J. C. (F1981). *Non-linear Evolution Equations.* Sound, colour, 12 min. [17]

Eilbeck, J. C. (F1982). *Kink Interactions in the ϕ^4 Model.* Silent, colour, 6 min.

Eilbeck, J. C. & Lomdahl, P. S. (F1982). *Sine–Gordon Solitons.* Sound, colour, 14 min.

The above films are also available on video. Enquiries about hiring or buying any of these should be made to Professor J. C. Eilbeck, Department of Mathematics, Heriot-Watt University, Riccarton, Edinburgh EH14 4AS, U.K.

Zabusky, N. J., Kruskal, M. D. & Deem, G. S. (F1965). *Formation, Propagation and Interaction of Solitons (Numerical Solutions of Differential Equations Describing Wave Motions in Nonlinear Dispersive Media).* Silent, b/w, 35 min. [17]

The above film may be loaned from Bell Telephone Laboratories, Inc. Film Library, Murray Hill, New Jersey 07971, U.S.A.

Subject index